Science and Technology Concepts–Secondary™

Experimenting
with **Forces**
and **Motion**

Student Guide

National Science Resources Center

The National Science Resources Center (NSRC) is operated by the Smithsonian Institution to improve the teaching of science in the nation's schools. The NSRC disseminates information about exemplary teaching resources, develops curriculum materials, and conducts outreach programs of leadership development and technical assistance to help school districts implement inquiry-centered science programs.

Smithsonian Institution

The Smithsonian Institution was created by an act of Congress in 1846 "for the increase and diffusion of knowledge..." This independent federal establishment is the world's largest museum complex and is responsible for public and scholarly activities, exhibitions, and research projects nationwide and overseas. Among the objectives of the Smithsonian is the application of its unique resources to enhance elementary and secondary education.

STC Program™ Project Sponsors

National Science Foundation

Bristol-Meyers Squibb Foundation

Dow Chemical Company

DuPont Company

Hewlett-Packard Company

The Robert Wood Johnson Foundation

Carolina Biological Supply Company

Science and Technology Concepts–Secondary™

Experimenting
with **Forces**
and **Motion**
Student Guide

The **STC** *Program*™

Smithsonian Institution
National Science Resources Center

www.carolinacurriculum.com

Published by Carolina Biological Supply Company
Burlington, North Carolina

NOTICE This material is based upon work supported by the National Science Foundation under Grant No. ESI-9618091. Any opinions, findings, and conclusions or recommendations expressed in this material are those of the authors and do not necessarily reflect views of the National Science Foundation or the Smithsonian Institution.

This project was supported, in part, by the **National Science Foundation**. Opinions expressed are those of the authors and not necessarily those of the foundation.

ISBN 978-1-4350-0697-3

Published by Carolina Biological Supply Company, 2700 York Road, Burlington, NC 27215.
Call toll free 1-800-334-5551.

Science and Technology Concepts—Secondary™
Experimenting with Forces and Motion

The following revision was based on the STC/MS™ module *Energy, Machines, and Motion.*

Developer
Dane J. Toler

Scientific Reviewer
Ramon E. Lopez
Professor of Physics
University of Texas at Arlington

Illustrator
Taina Litwak

Writer/Editor
Ana C. Jennings

Photo Research
Jane Martin
Devin Reese

National Science Resources Center Staff

Executive Director
Sally Goetz Shuler

Program Specialist/Revision Manager
Elizabeth Klemick

Contractor, Curriculum Research and Development
Devin Reese

Publications Graphics Specialist
Heidi M. Kupke

Carolina Biological Supply Company Staff

Director of Product and Development
Cindy Morgan

Product Manager, The STC Program™
Jack Ashton

Curriculum Editors
Lauren Goldsmith
Gary Metheny

Managing Editor, Curriculum Materials
Cindy Vines Bright

Publications Designers
Trey Foster
Charles Thacker
Greg Willette

Science and Technology Concepts for Middle Schools™
Energy, Machines, and Motion
Original Publication

Module Development Staff

Developer/Writer
Dane J. Toler

Science Advisor
John Layman
Professor Emeritus of Education and Physics
University of Maryland

Contributing Writers
Carolyn Hanson
Linda Harteker
David Marsland
Carol O'Donnell
Kitty Lou Smith
David Wetzel

Illustrators
Dan Sherbo
Max-Karl Winkler

STC/MS™ Project Staff

Principal Investigators
Douglas Lapp, Executive Director, NSRC
Sally Goetz Schuler, Deputy Director, NSRC

Project Director
Kitty Lou Smith

Curriculum Developers
David Marsland
Henry Milne
Carol O'Donnell
Dane J. Toler

Illustration Coordinator
Max-Karl Winkler

Photo Editor
Janice Campion

Graphic Designer
Heidi M. Kupke

STC/MS™ Project Advisors

Judy Barille, Chemistry Teacher, Fairfax County Public Schools, Virginia

Steve Christiansen, Science Instructional Specialist, Montgomery County Public Schools, Maryland

John Collette, Director of Scientific Affairs (retired), DuPont Company

Cristine Creange, Biology Teacher, Fairfax County Public Schools, Virginia

Robert DeHaan, Professor of Physiology, Emory University Medical School

Stan Doore, Meteorologist (retired), National Weather Service, National Oceanic and Atmospheric Administration

Ann Dorr, Earth Science Teacher (retired), Fairfax County Public Schools, Virginia; Board Member, Minerals Information Institute

Yvonne Forsberg, Physiologist, Howard Hughes Medical Center

John Gastineau, Physics Consultant, Vernier Corporation

Patricia A. Hagan, Science Project Specialist, Montgomery County Public Schools, Maryland

Alfred Hall, Staff Associate, Eisenhower Regional Consortium at Appalachian Educational Laboratory

Connie Hames, Geology Teacher, Stafford County Public Schools, Virginia

Jayne Hart, Professor of Biology, George Mason University

Michelle Kipke, Director, Forum on Adolescence, Institute of Medicine

John Layman, Professor Emeritus of Physics, University of Maryland

Thomas Liao, Professor and Chair, Department of Technology and Society, State University of New York at Stony Brook

Ian MacGregor, Director, Division of Earth Sciences, National Science Foundation

Ed Mathews, Physical Science Teacher, Fairfax County Public Schools, Virginia

Ted Maxwell, Geomorphologist, National Air and Space Museum, Smithsonian Institution

Tom O'Haver, Professor of Chemistry/Science Education, University of Maryland

Robert Ridky, Professor of Geology, University of Maryland

Mary Alice Robinson, Science Teacher, Stafford County Public Schools, Virginia

Bob Ryan, Chief Meteorologist, WRC Channel 4, Washington, D.C.

Michael John Tinnesand, Head, K–12 Science, American Chemical Society

Grant Woodwell, Professor of Geology, Mary Washington College

Thomas Wright, Geologist, National Museum of Natural History, Smithsonian Institution; U.S. Geological Survey (emeritus)

Acknowledgments

The National Science Resources Center gratefully acknowledges the following individuals and school systems for their assistance with the national field-testing of *Energy, Machines, and Motion:*

Atlanta Public School District, Atlanta, Georgia

Site Coordinator
Lela Blackburn, Science Coordinator

Lorrie D. Green, Teacher
J.C. Young Middle School

Stephanie Greene, Teacher
W.L. Parks Middle School

Melanie Robinson, Teacher
West Fulton Middle School

Hands-on Activity Science Program, Huntsville, Alabama

Site Coordinator
Sandra Enger
University of Alabama in Huntsville

Huntsville Area Public Schools
Nana Garner, Teacher
Athens Middle School

Jerry Lindsey, Teacher
Brookhaven Middle School

Sandra Patrick, Teacher
Scottsboro Junior High School

Judy Smith, Teacher
Brookhaven Middle School

Mason-Lake Oceana Mathematics and Science Center, Scottville, Michigan

Site Coordinator
Marsha Barter, Director

Mason-Lake Intermediate School District
Melissa Bansch, Teacher
O.J. DeJonge Junior High School

James R. Beckstrom, Teacher
O.J. DeJonge Junior High School

Mason County Central School District
Jean M. Nicholson, Teacher
Mason County Central Middle School

Montgomery County Public Schools, Montgomery County, Maryland

Site Coordinator
Patricia A. Hagan, Science Project Specialist

Tim O'Keefe, Teacher
Martin Luther King Middle School

**Schenectady City Schools,
Schenectady, New York**

Site Coordinator
Paul Scampini, Coordinator of Science

Site Coordinator
Ann Crotty, Professional
Development Specialist

Beverly Elander, Teacher
Oneida Middle School

Colette McCarthy, Teacher
Central Park Middle School

Fitz Glenn Miller
Mount Pleasant Middle School

**Fort Bend Independent School District,
Sugar Land, Texas**

Site Coordinator
Mary Ingle, Secondary Science Coordinator

Doug Fletcher, Teacher
Hodges Bend Middle School

Ruth McMahan, Teacher
Hodges Bend Middle School

Carl Peters, Teacher
Hodges Bend Middle School

The NSRC appreciates the contribution of its
STC/MS project evaluation consultants—

Program Evaluation Research Group (PERG), Lesley College

Sabra Lee
Researcher, PERG

George Hein
Director (retired), PERG

Center for the Study of Testing, Evaluation,
and Education Policy (CSTEEP), Boston College

Joseph Pedulla
Director, CSTEEP

Maryellen Harmon
Director (retired), CSTEEP

Preface

Community leaders and state and local school officials across the country are recognizing the need to implement science education programs consistent with the National Science Education Standards to attain the important national goal of scientific literacy for all students in the 21st century. The Standards present a bold vision of science education. They identify what students at various levels should know and be able to do. They also emphasize the importance of transforming the science curriculum to enable students to engage actively in scientific inquiry as a way to develop conceptual understanding as well as problem-solving skills.

The development of effective standards-based, inquiry-centered curriculum materials is a key step in achieving scientific literacy. The National Science Resources Center (NSRC) has responded to this challenge through Science and Technology Concepts–Secondary™. Prior to the development of these materials, there were very few science curriculum resources for secondary students that embodied scientific inquiry and hands-on learning. With the publication of STC–Secondary™, schools will have a rich set of curriculum resources to fill this need.

Since its founding in 1985, the NSRC has made many significant contributions to the goal of achieving scientific literacy for all students. In addition to developing Science and Technology Concepts–Elementary™—an inquiry-centered science curriculum for grades K through 6—the NSRC has been active in disseminating information on science teaching resources, preparing school district leaders to spearhead science education reform, and providing technical assistance to school districts. These programs have had a significant impact on science education throughout the country. The transformation of science education is a challenging task that will continue to require the kind of strategic thinking and insistence on excellence that the NSRC has demonstrated in all of its curriculum development and outreach programs. The Smithsonian Institution, our sponsoring organization, takes great pride in the publication of this exciting new science program for secondary students.

Letter to the Students

Smithsonian Institution
National Science Resources Center

Dear Student,

The National Science Resources Center's (NSRC) mission is to improve the learning and teaching of science for K–12 students. As an organization of the Smithsonian Institution, the NSRC is dedicated to the establishment of effective science programs for all students. To contribute to that goal, the NSRC has developed and published two comprehensive, research-based science curriculum programs: Science and Technology Concepts–Elementary™ and Science and Technology Concepts–Secondary™.

By using the STC–Secondary™ curriculum materials, we know that you will build an understanding of important concepts in life, earth, and physical sciences; learn critical-thinking skills; and develop positive attitudes toward science and technology. The National Science Education Standards state that all secondary students "...should be provided opportunities to engage in full and partial inquiries.... With an appropriate curriculum and adequate instruction, ... students can develop the skills of investigation and the understanding that scientific inquiry is guided by knowledge, observations, ideas, and questions."

STC-Secondary also addresses the national technology standards published by the International Technology Education Association. Informed by research and guided by standards, the design of the STC-Secondary units address four critical goals:

• Use of effective student and teacher assessment strategies to improve learning and teaching

• Integration of literacy into the learning of science by giving students the lens of language to focus and clarify their thinking and activities

• Enhanced learning using new technologies to help students visualize processes and relationships that are normally invisible or difficult to understand

• Incorporation of strategies to actively engage parents to support the learning process

We hope that by using the STC-Secondary curriculum you will expand your interest, curiosity, and understanding about the world around you. We welcome comments from students and teachers about their experiences with the STC-Secondary program materials.

Sally Goetz Shuler
Executive Director
National Science Resources Center

Navigating an STC–Secondary™ Student Guide

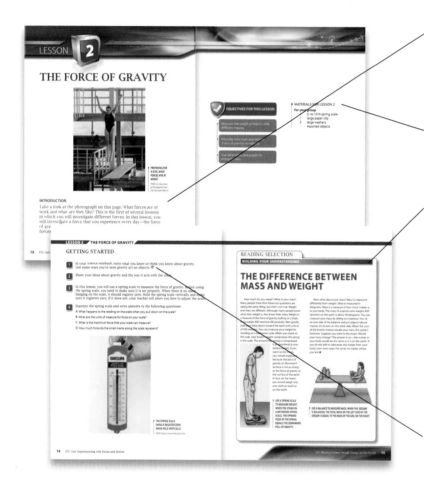

INTRODUCTION

This short paragraph helps get you interested about the upcoming inquiries.

MATERIALS

This helps you get organized and prepare for your inquiries.

READING SELECTION:
BUILDING YOUR UNDERSTANDING

These reading selections are part of the lesson, and give you information about the topic or concept you are exploring.

NOTEBOOK ICON

During the course of an inquiry, you'll record data in different ways. This icon lets you know to record in your science notebook. Student sheets are called out when you're to write there. You may go back and forth between your notebook and a student sheet. Watch carefully for the icon throughout the procedure.

SAFETY TIPS

Safety in the science classroom is very important. Tips throughout the student guide will help you to practice safe techniques while conducting investigations. It is very important to read and follow all safety tips.

SAFETY TIP

PROCEDURE

This tells you what to do. Sometimes the steps are very specific, and sometimes they guide you to come up with your own investigation and ways to record data.

REFLECTING ON WHAT YOU'VE DONE

These questions help you think about what you've learned during the lesson's inquiries, apply them to different situations, and generate new questions. Often you'll discuss your ideas with the class.

READING SELECTION: EXTENDING YOUR KNOWLEDGE

These reading selections come after the lesson, and show new ways that the topic or concept you learned about during the lesson can be applied, often in real-world situations.

GLOSSARY

Here you can find scientific terms defined.

INDEX

Locate specific information within the student guide using the index.

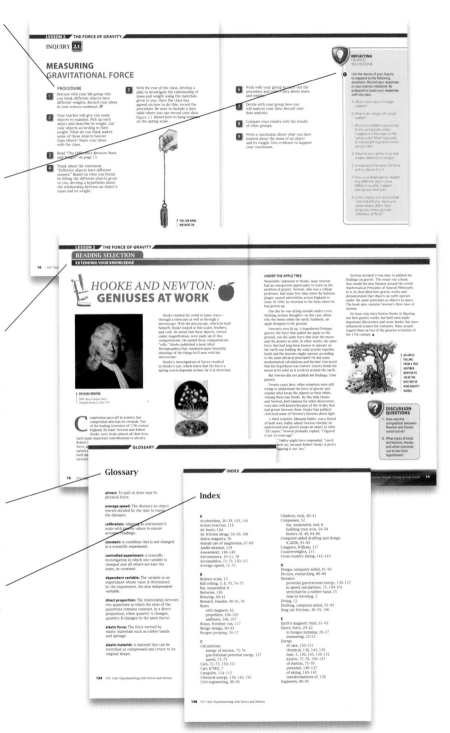

Contents

CONTENTS

CIRCUIT OF INQUIRIES— A PRE-ASSESSMENT

INTRODUCTION

What do you know about forces, energy, and motion? In this lesson, you will complete six short inquiries about the topics in this unit. The inquiries are designed to get you thinking about physical forces, energy transformations, and the motion of objects. The observations you make and the ideas you discuss in this lesson will prepare you for future inquiries in *Experimenting with Forces and Motion*.

CAN YOU SEE MOTION IN THIS PHOTO? CAN YOU SEE FORCES BEING APPLIED? CAN YOU SEE EVIDENCE OF ENERGY? YOU WILL EXPLORE THESE TOPICS IN THIS UNIT.

PHOTO: © Terry G. McCrea/ Smithsonian Institution

OBJECTIVES FOR THIS LESSON

Perform a series of activities to investigate forces, energy transformations, and motion.

Observe, describe, and hypothesize about the physical phenomena you experiment with in the activities.

Relate your observations to personal experiences.

For you

1   copy of Student Sheet 1: What We Observe About Energy, Forces, and Motion

GETTING STARTED

1 Your teacher will divide the class into groups of three. You and your lab partners will work together on the six short inquiries in this lesson.

2 Listen as your teacher reviews the directions for carrying out the inquiries. At your teacher's direction, go with your group to your first inquiry station.

PROCEDURE FOR THE CIRCUIT OF INQUIRES

1 At each station, complete the activity described in the directions on the Inquiry Card. The directions are also included on pages 5–8 in your Student Guide. You may want to refer to them later.

2 For each activity, record your observations on Student Sheet 1: What We Observe About Energy, Forces, and Motion. Answer the questions in complete sentences.

3 When you finish each inquiry, put everything back the way you found it for the next group of students.

4 When time is called, move quickly and quietly to the next station. Repeat Steps 1, 2, and 3 for the other five inquiries.

5 Complete your student sheet and make sure you have used complete sentences.

▶ **WHAT FORCES ARE ABOUT TO ACT ON THIS BALL?**

PHOTO: chispita_666/
creativecommons.org

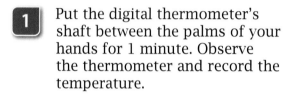

INQUIRY 1.1

THE HAND WARMER

PROCEDURE

1 Put the digital thermometer's shaft between the palms of your hands for 1 minute. Observe the thermometer and record the temperature.

SAFETY TIP

When using a digital thermometer, be careful with the pointed end.

2 Leaving the thermometer between your hands, rub your hands together for a short time. Observe the thermometer as you do this. Stop rubbing your hands and describe what happens to the temperature.

3 Write down another example of how you can produce heat.

4 Each student should perform Steps 1 and 2 if there is time.

INQUIRY 1.2

CONSTRUCTING A GRAPH

PROCEDURE

1 A bowling ball was released, and the distance it rolled was measured at the end of 5, 10, and 15 seconds.

TABLE 1 DISTANCE THE BOWLING BALL TRAVELED OVER TIME

TIME (SECONDS)	DISTANCE (METERS)
5	15
10	25
15	30

2 Draw a graph using the measurements in Table 1.

3 What does the graph show about the bowling ball's motion?

INQUIRY 1.3

ROLLING AROUND

PROCEDURE

1 Place the steel ball inside the roll of tape, then use your hand to move the roll of tape in a circular motion to make the ball move around the inside of the roll of tape (see Figure 1.1). When the ball is rolling steadily, quickly lift up the roll of tape and watch how the ball moves.

2 Do you think there are any forces acting on the ball when it is rolling around the inside of the roll of tape? Why or why not?

3 Write a description of the motion of the ball after the roll of tape is lifted. Do you think there are any forces acting on the ball after the roll of tape is lifted? Why or why not?

▶ **SETUP FOR INQUIRY 1.3: ROLLING AROUND**
FIGURE **1.1**

INQUIRY 1.4

THE PUCK LAUNCHER

PROCEDURE

1 Put the puck against the rubber band as shown in Figure 1.2. Pull the band and puck back about 2 centimeters (cm) and release the puck. Describe the puck's motion.

2 Put the puck against the rubber band. Pull the band and puck back about 4 cm and release the puck. Describe the puck's motion.

3 What force acted on the puck each time it was released?

4 Describe how the puck's motion is different and how it is the same in the two trials.

5 Give an example of a force that acts on an object. What is the force's effect?

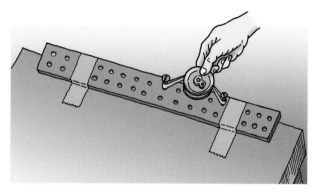

▶ **SETUP FOR THE PUCK LAUNCHER**
FIGURE **1.2**

INQUIRY 1.5

DOWN THE RAMP

PROCEDURE

1 Put the car on the ramp's high end (Position 1) and let it go.

2 Describe the car's motion.

▶ **THE CAR AND THE RAMP**
FIGURE **1.3**

3 Put the car farther down the ramp (Position 2) and let it go (see Figure 1.3).

4 Describe the motion you observe.

5 Compare the motions of the two cars and explain why they are different.

INQUIRY **1.6**

THE SUSPENDED BAR

PROCEDURE

1 Bring Bar 1 on the table near the suspended bar (see Figure 1.4). Describe what happens.

2 Turn Bar 1 around and bring it near the suspended bar. What happens?

3 Take the compass on the table and place it near the suspended bar. Move it around the bar. What happens to the compass needle?

4 What forces are acting here?

5 Can you make the suspended bar spin without touching it? What kind(s) of energy can you associate with the spinning bar?

REFLECTING
ON WHAT
YOU'VE DONE

1 Discuss with your partner and then with the class what you have observed and your ideas about your observations.

2 In class, summarize your observations and give examples of similar forces, energy transformations, and motions you have seen outside the classroom. Your teacher will record the ideas on a class list.

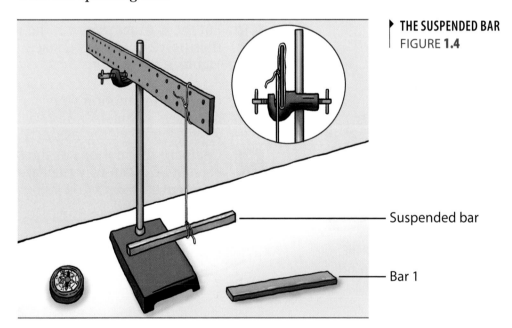

▶ **THE SUSPENDED BAR**
  FIGURE **1.4**

———— Suspended bar

———— Bar 1

Galileo and EXPERIMENTAL SCIENCE

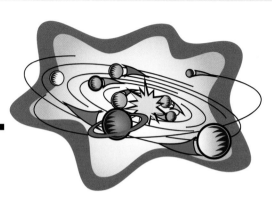

▶ **GALILEO GALILEI**
PHOTO: Library of Congress, Prints & Photographs Division, LC-USZ62-47604

n Lesson 1, you made observations, took measurements, recorded and analyzed data, and discussed your findings with your classmates. This probably seems like the logical way to do science—but it hasn't always been the case.

In ancient times, scientists were curious about the world around them. These early scientists often relied too much on general observations and on what previous scientists had done. They were often reluctant to question authority.

About 400 years ago in Western Europe, things changed. A number of scientists began to explore the world around them with a fresh eye. Everything interested them. They looked at things in a new way. They did not just observe things and record information; they experimented to see if their ideas were correct. One of the most famous of these scientists was an Italian named Galileo Galilei.

Galileo was born in Pisa, Italy, in 1564. At the age of 17, he entered the University of Pisa. He planned to become a doctor, but he soon became sidetracked. Galileo began to observe things that were happening around him, and he found

READING SELECTION
EXTENDING YOUR KNOWLEDGE

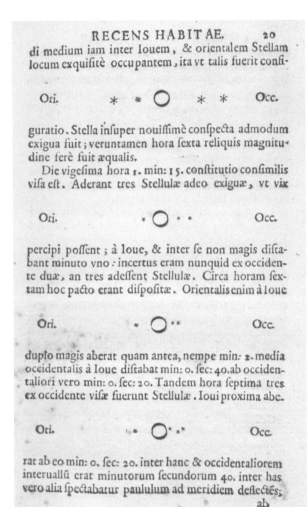

them much more interesting than what he heard in the lecture hall.

Even the simplest things could be fascinating. For example, Galileo sat in church and watched a lamp swing from the ceiling. He soon realized that its movements were regular. He could time them with his pulse beat. When Galileo watched different lamps, he discovered that there was a relationship between the time it took for a lamp to swing back and forth and the length of the chain from which it was suspended. He also discovered that a lamp swung back and forth in the same amount of time, regardless of how broad or narrow the swing.

Galileo did not find any immediate application for his observations of the swinging lamp; that would come later, with the invention

WITH HIS TELESCOPE, GALILEO SAW FOUR MOONS ORBITING JUPITER. HE STUDIED THEIR MOTIONS FROM NIGHT TO NIGHT AND RECORDED THE POSITIONS OF THE MOONS IN HIS NOTEBOOK.

PHOTO: Courtesy of Smithsonian Institution Libraries, Dibner Library of the History of Science and Technology, Washington, D.C.

CLOSE-UP PHOTOS OF JUPITER'S MOONS, TAKEN BY THE *GALILEO* SPACE PROBE.

PHOTO: NASA Jet Propulsion Laboratory

of a pendulum clock. But it didn't matter. The experience was important because he had identified and documented a mathematical relationship in a universal event—the swinging of a lamp.

As Galileo became more involved in science, he began to record his observations in notebooks. This was another important distinction between him and earlier scientists. These notebooks, in which he frequently made sketches, enabled Galileo to share his ideas with other people of his time. The notebooks, which still exist today, give us insight into his imaginative and creative mind.

Galileo was also a famous inventor. One of his most astounding devices was a military compass that could aim cannonballs at the enemy. He achieved his greatest fame as an astronomer, however. He built his own telescope. With it, he made observations that revolutionized our understanding of the universe. He saw craters on the moon and thousands of stars in the Milky Way galaxy.

In 1609, Galileo looked at the planet Jupiter and saw four small points of light circling it. At first, he thought they were distant stars. As he continued to observe and record what he saw, he finally concluded that those points of light were actually moons in orbit around the planet. Today, these moons are called the Galilean moons.

Galileo's ideas sometimes got him into trouble. For example, his observations convinced him that the planet Earth revolves around the sun. (For centuries, people had thought that Earth was the center of the solar system.) This idea was very controversial, especially to the leaders of the Church, who put Galileo on trial for heresy and threatened him with torture. To keep Galileo quiet, the Church leaders put him under house arrest for the rest of his life. Galileo could no longer speak in public, but he remained convinced that his beliefs about the solar system were correct (and, of course, they were).

The *Galileo* space probe, launched by the National Aeronautics and Space Administration (NASA) in 1989, honored this famous Italian scientist. Its mission was to observe Jupiter and send information back to Earth. The space probe also sent back information about the four moons that Galileo saw.

To Galileo, these moons were four tiny points of light. Take a look at the pictures from the space probe on page 10. Do you think Galileo would be pleased to see his moons in such detail? ∎

DISCUSSION QUESTIONS

1. What process did Galileo use to come to his conclusions?

2. What other famous scientists have gotten in trouble for their ideas? Why?

THE FORCE OF GRAVITY

▶ **PREPARING FOR A DIVE. WHAT FORCES ARE AT WORK?**

PHOTO: U.S. Navy photo by Photographer's Mate 2nd Class Isaiah Sellers III

INTRODUCTION

Take a look at the photograph on this page. What forces are at work and what are they like? This is the first of several lessons in which you will investigate different forces. In this lesson, you will investigate a force that you experience every day—the force of gravity. In the lessons that follow you will investigate other forces, including friction, elastic forces, and magnetic forces.

 OBJECTIVES FOR THIS LESSON

Measure the weight of objects with different masses.

Describe how mass and weight (force of gravity) are related.

Use data tables and graphs to interpret data.

▶ **MATERIALS FOR LESSON 2**

For your group

1	0- to 10-N spring scale
1	large paper clip
5	large washers
	Assorted objects

GETTING STARTED

1 In your science notebook, write what you know or think you know about gravity. List some ways you've seen gravity act on objects. 📝

2 Share your ideas about gravity and the way it acts with the class.

3 In this lesson, you will use a spring scale to measure the force of gravity. Before using the spring scale, you need to make sure it is set properly. When there is no mass hanging on the scale, it should register zero. Hold the spring scale vertically and make sure it registers zero. If it does not, your teacher will show you how to adjust the scale.

4 Examine the spring scale and write answers to the following questions:

A. What happens to the reading on the scale when you pull down on the scale?

B. What are the units of measure for force on your scale?

C. What is the maximum force that your scale can measure?

D. How much force do the small marks along the scale represent?

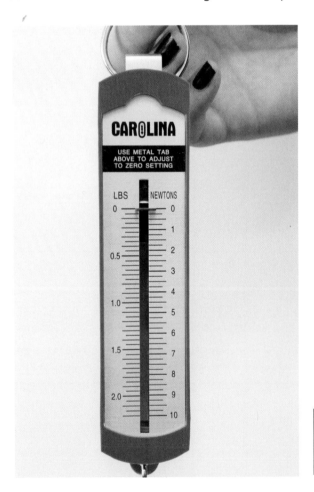

▶ THE SPRING SCALE
SHOULD REGISTER ZERO
WHEN HELD VERTICALLY.

PHOTO: National Science Resources Center

THE DIFFERENCE BETWEEN MASS AND WEIGHT

How much do you weigh? What is your mass? Many people think that these two questions are asking the same thing, but that's not true. Weight and mass are different. Although many people know what their weight is, few know their mass. Weight is a measure of the force of gravity pulling on a body. If you weigh 400 newtons (90 pounds), then gravity pulls your body down toward the earth with a force of 400 newtons. You can measure your weight by standing on a bathroom scale. When you stand on the scale, your body's weight compresses the spring in the scale. The amount the spring is compressed is proportional to your body's weight. If you went to the moon, you would weigh less because the force of gravity on the moon's surface is not as strong as the force of gravity on the surface of the earth. In fact, on the moon, you would weigh only one-sixth as much as on the earth.

▶ USE A SPRING SCALE TO MEASURE WEIGHT. WHEN YOU STAND ON A BATHROOM SPRING SCALE, THE UPWARD PUSH OF THE SPRING EQUALS THE DOWNWARD PULL OF GRAVITY.

Now what about your mass? Mass is measured differently from weight. Mass is measured in kilograms. Mass is a measure of how much matter is in your body. The mass of a person who weighs 400 newtons on the earth is about 40 kilograms. You can measure your mass by sitting on a balance. You sit on one side of the balance and put objects whose masses are known on the other side. When the sum of the known masses equals your mass, the system balances. Suppose you went to the moon. Would your mass change? The answer is no—the matter in your body would be the same as it is on the earth. If you do not add or take away any matter from your body, your mass stays the same, no matter where you are. ■

▶ USE A BALANCE TO MEASURE MASS. WHEN THE SEESAW IS BALANCED, THE TOTAL MASS ON THE LEFT SIDE OF THE SEESAW IS EQUAL TO THE MASS OF THE GIRL ON THE RIGHT.

INQUIRY **2.1**

MEASURING GRAVITATIONAL FORCE

PROCEDURE

1 Discuss with your lab group why you think different objects have different weights. Record your ideas in your science notebook. 🖉

2 Your teacher will give you some objects to examine. Pick up each object and describe its weight. List your objects according to their weight. What do you think makes some of these objects heavier than others? Share your ideas with the class.

3 Read "The Difference Between Mass and Weight" on page 15.

4 Think about the statement, "Different objects have different masses." Based on what you found by lifting the different objects given to you, develop a hypothesis about the relationship between an object's mass and its weight.

5 With the rest of the class, develop a plan to investigate the relationship of mass and weight using the materials given to you. Once the class has agreed on how to do this, record the procedure. Be sure to include a data table where you can record your data. Figure 2.1 shows how to hang washers on the spring scale.

▶ **YOU CAN HANG WASHERS ON THE HOOK OF THE SPRING SCALE.**
FIGURE **2.1**

6   Work with your group to carry out the procedure and collect data about mass and weight.

7   Decide with your group how you will analyze your data. Record your data analysis.

8   Compare your results with the results of other groups.

9   Write a conclusion about what you have learned about the mass of an object and its weight. Give evidence to support your conclusion.

REFLECTING
ON WHAT
YOU'VE DONE

1   Use the results of your inquiry to respond to the following questions. Record your responses in your science notebook. Be prepared to share your responses with the class.

A. What is the mass of a single washer?

B. What is the weight of a single washer?

C. When you added more washers to the spring scale, what happened to the mass on the spring scale? What happened to the weight registered on the spring scale?

D. What do you call the force that makes objects have weight?

E. In what direction does this force pull on the washers?

F. From your observations, explain why different objects have different weights. Support your answer with data.

G. In this inquiry you investigated one kind of force. From your observations of the force of gravity, write a general definition of "force."

HOOKE AND NEWTON: GENIUSES AT WORK

▶ **SIR ISAAC NEWTON**

PHOTO: Library of Congress, Prints &
Photographs Division, LC-USZ62-10191

Hooke studied his world in many ways—through a telescope as well as through a microscope. With the microscope, which he built himself, Hooke looked at fish scales, feathers, and cork. He noted that these objects, viewed under magnification, were made up of tiny compartments. He named these compartments "cells." Hooke published a book titled *Micrographica* that contained many beautiful drawings of the things he'd seen with his microscope.

Hooke's investigations of forces resulted in Hooke's Law, which states that the force a spring exerts depends on how far it is stretched.

▶ **THESE MAY LOOK LIKE FLOWERS IN A FIELD, BUT THEY ARE ACTUALLY MOLD SPORES AS SEEN THROUGH HOOKE'S MICROSCOPE.**

PHOTO: Library of Congress, Prints & Photographs Division, LC-USZ62-44624

Cooperation pays off in science, but competition also has its rewards. Two of the leading scientists of 17th century England, Sir Isaac Newton and Robert Hooke, were rivals almost all their lives. Each made important contributions to physics. Robert Hooke discovered the nature of elastic force, and Sir Isaac Newton investigated light. The nature of gravity was a subject that fascinated both men, but they also had many other scientific interests.

UNDER THE APPLE TREE

Meanwhile, unknown to Hooke, Isaac Newton had an unexpected opportunity to work on the problem of gravity. Newton, who was a college professor, had some free time when the bubonic plague caused universities across England to close. In 1666, he returned to the farm where he had grown up.

One day he was sitting outside under a tree, thinking serious thoughts—in this case, about why the moon orbits the earth. Suddenly, an apple dropped to the ground.

Newton's eyes lit up. A hypothesis! Perhaps gravity, the force that pulled the apple to the ground, was the same force that kept the moon and the planets in orbit. In other words, the same force that had long been known to operate on the earth was holding the solar system together. Earth and the heavens might operate according to the same physical principles! He did some mathematical calculations and became convinced that his hypothesis was correct: Gravity holds the moon in its orbit as it revolves around the earth.

But Newton did not publish his findings. Time passed.

Twenty years later, other scientists were still trying to understand the force of gravity and explain what keeps the planets in their orbits. Among them was Hooke. By this time Hooke and Newton, both famous for other discoveries, were also well known because of the rivalry that had grown between them. Hooke had publicly criticized some of Newton's theories about light.

A third scientist, Edmund Halley, was a friend of both men. Halley asked Newton whether he understood how gravity keeps an object in orbit. "Of course," Newton probably replied, "I figured it out 20 years ago."

"Well," Halley might have responded, "you'd better watch out, because Robert Hooke is pretty close to figuring it out, too."

Newton decided it was time to publish his findings on gravity. The result was a book that would become famous around the world: *Mathematical Principles of Natural Philosophy.* In it, he described how gravity works and demonstrated that objects on earth operate under the same principles as objects in space. The book also contains Newton's three laws of motion.

Sir Isaac may have beaten Hooke in figuring out how gravity works, but both men made important discoveries and wrote books that have influenced science for centuries. Many people regard them as two of the greatest scientists of the 17th century. ∎

▶ AN APPLE FALLING FROM A TREE INSPIRED NEWTON TO SOLVE THE MYSTERY OF HOW GRAVITY WORKS.

DISCUSSION QUESTIONS

1. How was the competition between Newton and Hooke constructive?

2. What types of tools did Newton, Hooke, and other scientists use to test their hypotheses?

THE FORCE OF A RUBBER BAND

▶ HAVING FUN ON A TRAMPOLINE. WHAT FORCES ARE AT WORK?

PHOTO: Wikimedia Commons

INTRODUCTION

Trampolines are a lot of fun. What forces are at work as you bounce up and down on a trampoline? You just investigated one of these forces, gravity, in Lesson 2. In this lesson you will investigate another force that also makes it possible to have so much fun.

OBJECTIVES FOR THIS LESSON

Describe the nature of elastic forces and how they act on objects.

Determine how the force a rubber band exerts is related to its stretch.

Use data tables and graphs to interpret data.

▶ **MATERIALS FOR LESSON 3**

For you

1	copy of Student Sheet 3.1: What Is the Elastic Force of the Rubber Band?

For your group

1	0- to 10-N spring scale
4	rubber bands
2	metersticks
1	piece of masking tape

GETTING STARTED

1. In your science notebook, write what you know or think you know about forces that are stretchy or springy. List some places where you see these kinds of forces in action. ✐

2. Share your ideas about these forces and where you see them at work with the class.

3. In this lesson you will use a spring scale to measure the force of a rubber band, but you will hold the spring scale horizontally instead of vertically. Before using the spring scale, make sure it is set properly. When the spring scale is horizontal and not pulling on anything, it should register zero. To check your spring scale, lay the spring scale with nothing attached to it on the table and make sure it registers zero. If it does not, adjust the scale so that it registers zero with no force pulling on it.

▶ **HOW COULD YOU DETERMINE THE ELASTIC FORCE OF A RUBBER BAND?**

PHOTO: Mykl Roventine/
creativecommons.org

MEASURING ELASTIC FORCE

PROCEDURE

1 Hold a rubber band at one end, and with a finger of your other hand, carefully stretch the rubber band far enough to feel the force of the pull. After doing this a few times, answer the following questions in your science notebook: ✎

A. When you stretch the rubber band, what do you feel?

B. In what direction does the rubber band pull your finger?

2 Repeat Step 1, but pull a little harder on the rubber band. Answer the following questions:

A. Now what do you feel?

B. What happens to the rubber band when you pull a little harder?

3 Discuss this question with your group: How could you find out how the force of the rubber band changes as the band is stretched?

4 Here's one way to answer the question in Step 3: Place one meterstick along the edge of your lab table. Slide a rubber band over the meterstick and move it to the middle of the meterstick, as shown in Figure 3.1.

▸ MOVE THE RUBBER BAND TO THE MIDDLE OF THE METERSTICK.
FIGURE **3.1**

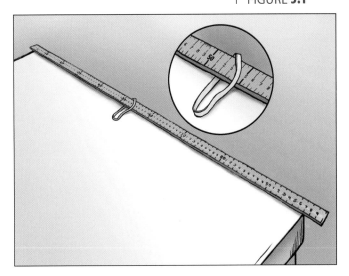

Inquiry 3.1 continued

5 Now place the other meterstick on the table so it is perpendicular to the meterstick along the edge of the table and next to the rubber band, as shown in Figure 3.2. You may want to tape this meterstick to the table so that it will not slip or move.

6 Hook the spring scale to the free end of the rubber band. The rubber band should be straight but not stretched, as shown in Figure 3.3.

7 Have two of your group members hold the meterstick along the edge of the table while another group member pulls slowly on the spring scale until the rubber band is stretched 2.0 cm. Observe and record the spring scale's force reading in Table 1 on Student Sheet 3.1: What Is the Elastic Force of the Rubber Band?

8 Stretch the rubber band another 2.0 cm and record the force. Repeat this process for each 2.0-cm interval shown in the data table. Every time you stretch the rubber band, record the force on the spring scale.

▶ MAKE SURE THE OTHER METERSTICK IS PERPENDICULAR TO THE ONE ON THE EDGE OF THE TABLE.
FIGURE **3.2**

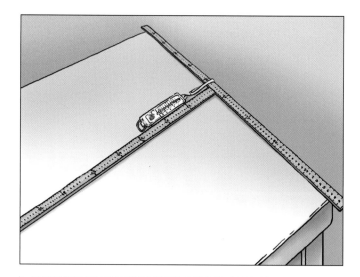

▶ HOOK THE SPRING SCALE TO THE RUBBER BAND SO THE BAND IS STRAIGHT BUT NOT STRETCHED.
FIGURE **3.3**

9 Look at the measurements in your data table and answer Question 1 on Student Sheet 3.1: What happens to the force required to stretch the rubber band when the stretching distance is doubled? Review at least three instances of doubling the distance.

10 Use the data you collected to make a graph. Remember to plot the dependent variable versus the independent variable. Make sure both axes are uniformly scaled and properly labeled. Give your graph a title.

11 Use your graph to answer Questions 3, 4, and 5 on Student Sheet 3.1:

- What does your graph tell you about the relationship between the force needed to stretch the rubber band and the distance it stretches?

- Did the force needed to stretch the rubber band increase by the same amount each time you stretched the rubber band another 2.0 cm? Is the answer visible in the graph?

- Can you predict the force needed to stretch the rubber band to 40.0 cm? Why or why not?

12 Discuss your findings with the class.

REFLECTING
ON WHAT
YOU'VE DONE

1 In your notebook, write your responses to the following questions. Be prepared to share your responses with the class.

A. You want to pull a cart along the floor. How would you use the rubber band to do this?

B. Suppose you want to apply a force twice as big as the one you would use for Question 1. What would you do to the rubber band to produce twice as much force?

C. How is the spring scale you used to measure force like the rubber band?

D. In this inquiry, you examined forces that pull on objects. Give an example of another way to exert a force on an object.

E. Materials like rubber bands and springs are called "elastic" materials. Can you think of other materials that are elastic?

F. Review your definition of force that you wrote in Lesson 2. How would you now define "force"? Write a new definition.

BUNGEE JUMPING:
THE FORCES
ARE WITH YOU

Bungee jumping isn't for wimps! You leap from a bridge or high platform and drop like a rock for hundreds of feet. And just when you're thinking that you'll never stop falling, you do. You stop just short of hitting the ground. What saved you? You may think it's the thin elastic cord around your ankles. You just experienced two powerful forces—gravity and elastic force.

REAL SWINGERS!

You might think that bungee jumping is as new as the closest amusement park, but actually it's been around for centuries. In a village called Bunlap on Pentecost Island in the South Pacific Ocean, it is an annual ritual. The islanders call it land diving.

Each year at harvest time, the men of the village erect towers made of tree trunks, branches, and vines. These towers are 15 to 25 meters (49.2 to 82 feet) high. Then each man selects some of the same vines to make his own bungee cord. One after another, the men and older boys take their homemade bungee cords and climb the tower. The younger jumpers climb only part of the way, but the older men go to the very top. Before the jump, friends tie a vine to each of the jumper's legs. They anchor the other end to the tower.

The jumper takes his position. Below, the villagers sing and dance. The jumper raises his arm, and the crowd becomes silent. The jumper plunges to earth. If all goes as planned, the springy, elastic vines break his fall just before he reaches the ground. Fellow members of the tribe rush to untie the vines and begin to dance in his honor.

▶ A BUNGEE JUMPER TAKES THE PLUNGE OFF A PLATFORM IN SHI DU, CHINA. GRAVITY AND ELASTIC FORCE WORK TOGETHER IN THIS THRILLING SPORT.

PHOTO: Russell Neches/creativecommons.org

JUMPING ACROSS CONTINENTS

Bungee jumping became popular in the western world in 1979 when members of the Dangerous Sports Club at Oxford University, who had read about the land divers of Pentecost Island, leaped from the 75-meter-high (246-foot-high) Clifton Bridge in the city of Bristol, England. Because it was an important occasion, they were all wearing tuxedos and top hats! Later, these men traveled around the world. They leaped from the Golden Gate Bridge in California and the Royal Gorge in Colorado.

As soon as people heard of this thrilling new sport, many wanted to try it for themselves. Soon they were lining up at amusement parks and paying for the privilege of taking a death-defying leap.

THERE'S SCIENCE BEHIND IT

Suppose you want to make a jump. How would you convince your mom or dad that it's really not dangerous?

You'd tell them that when you're bungee jumping, the forces are with you—the force of gravity and elastic force, that is.

Let's take an imaginary jump and see what happens.

First, the operator hooks you up to a bungee cord—a stretchy, elastic material. After a few instructions, you're ready to leap. You close your eyes and step off into thin air. With nothing to hold you up, the force of gravity accelerates you downward.

You open your eyes. The ground is rushing toward you like a railroad train. As you near the ground, the bungee cord stretches like a gigantic rubber band. The more the cord stretches, the stronger its force becomes.

Finally, you reach the point where the upward pull of the cord is greater than the downward pull of gravity. You feel an upward pull. ("Thank you, elastic force!") Your fall slows then stops—inches above the ground. ("Thank you, thank you!")

It's not over yet. The upward pull of the cord, now stronger than gravity, slings you back into the air. You fall again. Rise and fall, rise and fall—you feel like a fish at the end of a line! Each time, the movement becomes less extreme. You don't even notice by now. What matters is that you *survived*.

And if the line is not too long, you'll have time for one more jump before the park closes! ■

▶ BOTH MEN AND BOYS ON PENTECOST ISLAND PARTICIPATE IN LAND DIVING. THEY TIE BUNGEE CORDS MADE FROM SPRINGY LIANA VINES AROUND THEIR ANKLES AND LEAP FROM THE TOP OF TOWERS 25 METERS HIGH.

PHOTO: © Mark Karrass/Corbis

▶ A DIVER TAKES THE PLUNGE FROM THE MAST OF A TALL SHIP.

PHOTO: lpiepiora/
creativecommons.org

DISCUSSION QUESTIONS

1. What factors have to be taken into consideration when a person wants to go bungee jumping? Does their height or weight matter? What about the equipment?

2. In what other ways do people use elastic forces?

THE FORCE OF FRICTION

▶ **WHAT FORCE STOPS THE BASEBALL PLAYER AS HE SLIDES INTO THE BASE?**

PHOTO: jmagnusphoto/
creativecommons.org

INTRODUCTION

In Lessons 2 and 3, you investigated the properties of gravity and elastic forces in rubber bands. In this lesson, you will explore another force—friction. Friction is another force you experience every day. People spend time, money, and effort to reduce it. You might be surprised to learn that friction is not always bad. It can be very helpful.

You will investigate the force of friction as you pull a wooden block at a constant speed across a surface. As you do so, you will investigate three variables—surface type, weight (load), and surface area (base area)—and determine how each affects the frictional force between the block and the surface. In each trial, you will change one variable and keep the other two the same. This method is an example of good experimental design. As in the previous inquiries, you will record observations, collect data, and draw conclusions based on what you discover about friction. Understanding friction will help you understand motion, which you will study later in this unit.

OBJECTIVES FOR THIS LESSON

- Observe the properties of sliding friction.

- Measure the force of friction on a wooden block pulled across different surfaces.

- Measure the force of friction on loads of different weights.

- Measure the force of friction on a wooden block with different base areas in contact with a surface.

▶ **MATERIALS FOR LESSON 4**

For you

1	copy of Student Sheet 4.1: What a Drag!
1	copy of Student Sheet 4.2: Changing the Load

For your group

2	wooden blocks with attached screw hook
1	0- to 2.5-N spring scale
1	0- to 10-N spring scale
1	piece of waxed paper
1	piece of paper towel
1	piece of fine sandpaper
1	piece of coarse sandpaper
1	meterstick
1	rubber band
1	piece of masking tape

GETTING STARTED

1 Discuss what you know about friction with your group and then with the rest of the class.

2 Identify a situation in which friction works against you. Then identify one in which friction works for you. Think of situations where friction is very low or almost zero. What would happen if there was no friction at all?

3 Your teacher will share information with you about how to use a spring scale to measure the force of friction. Listen carefully and ask questions if you are not sure how to make the measurements.

▸ **WHAT EVIDENCE OF FRICTION DO YOU SEE IN THIS PHOTO?**

PHOTO: National Science Resources Center

PULLING A BLOCK ACROSS DIFFERENT SURFACES

PROCEDURE

1 Lay a spring scale on the tabletop. Does the pointer register zero? If not, adjust it to read zero.

2 In this inquiry, you will investigate the force of friction on a block as you pull it across each of the following surfaces: the plain tabletop, waxed paper, a paper towel, fine sandpaper, and coarse sandpaper. Tape the four surface strips (waxed paper, paper towel, fine sandpaper, coarse sandpaper) to the tabletop. Measure and mark off another 27-cm length on the bare tabletop.

3 You will need to decide which spring scale to use for your measurements. If you are not sure which to use, try making some measurements of the forces needed to pull the block using the scale. The best scale to use is one that will register the force to pull the block, yet not go off scale when you exert the maximum force needed to pull the block.

4 Attach the hook on the spring scale to the round screw hook on the wooden block, as shown in Figure 4.1.

5 How do you think the frictional force on the block will compare as you pull the block across the different surfaces? Write your prediction in your science notebook. 🖉

▸ **WOODEN BLOCK CONNECTED TO A SPRING SCALE**
FIGURE **4.1**
PHOTO: ©2009 Carolina Biological Supply Company

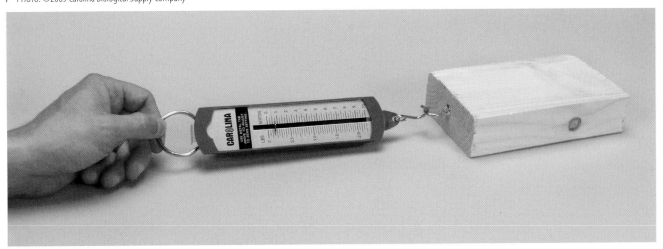

Inquiry 4.1 continued

6 Before you start to collect data, practice pulling the wooden block with the spring scale across a surface at a steady rate, as shown in Figure 4.2. Make sure you hold the spring scale parallel to the tabletop and pull horizontally on the block. As you move the block at a steady speed, observe the force reading. You will probably find that the force is not perfectly steady. When the spring scale reading is not steady, it is best to do several trials and average your results. How many trials should you do for each surface? Discuss this with your group and decide on the number of trials needed in order to obtain accurate force data.

7 Pull the block across each of the five surfaces and collect force data for each surface. Be sure to pull the block across the entire length of the surface. Record your data in Table 1 on Student Sheet 4.1: What a Drag! Calculate the average force for each surface.

8 How can you graphically represent the average force data for each surface? What kind of graph should you use? Construct a graph.

9 In your science notebook, record your answers to the following questions: 🖉

A. Which surface required the greatest force to pull the wooden block across it?

B. Which surface required the least force?

C. Did the weight of your wooden block change as the surfaces changed?

D. Review the variables for this lesson. Which variables did not change as you tested each surface? Be prepared to discuss your answers with the class.

▸ **PULLING A WOODEN BLOCK ACROSS A SURFACE TO MEASURE THE FORCE OF FRICTION**
FIGURE **4.2**
PHOTO: ©2009 Carolina Biological Supply Company

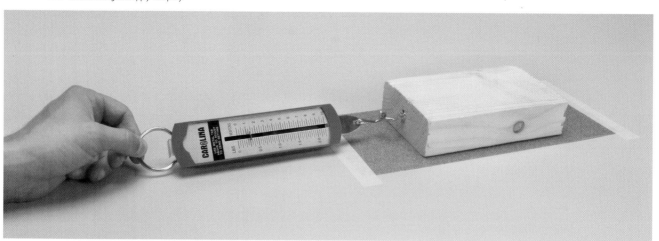

CHANGING THE LOAD

PROCEDURE

1 Think about and discuss the following question with your lab group: What will happen to the effort force needed to pull the block if you change the weight of the block? Write your prediction in your science notebook. 🖉

2 This inquiry requires that you and your lab group work as a team and collaborate with the other teams in your class. There is not enough time for each individual team to check all surfaces with different loads. Therefore, you will need to work together as a class and assign different surfaces to different teams. Each team needs to collect data. Each surface should be assigned to at least two teams.

3 Design a data table to record your measurements on Student Sheet 4.2: Changing the Load. Discuss with the class how you will measure the weight of the load in each trial.

4 Change the weight of the load by stacking more blocks, one at a time, on the original block. Each time you add a block, measure the force it takes to pull this load at a steady speed across the surface. Share blocks with another team so both teams can gather data for a load of up to four blocks.

5 Graph your data using the grid on Student Sheet 4.2.

6 Compare your data and graph with that of a team that used the same surface. Discuss with your group and the other team the relationship between the load (total weight of the block or blocks) and the frictional force. Record the description in your science notebook. 🖉

7 Compare your data with the data of groups that used different surfaces. What do you find? Record your findings.

8 Think about the variables in this investigation. What did you keep constant as you changed the weight of the blocks? Record your answer, and then share your results with the class.

INQUIRY 43

CHANGING THE SURFACE AREA

PROCEDURE

1 Look at your block. You can turn the block on its wide side and pull on it, or stand it on one of its narrow sides and pull on it. When the block is on its wide side, the area in contact with the surface is greater than when it is on its narrow side. You pulled the block across the surfaces on its wide side in previous trials. Predict what will happen to the force of friction if you pull the block on a narrow side across each surface. Write your prediction in your science notebook.

2 Construct a data table to record the description of each surface area of the block (wide or narrow) and the measurement of the force needed to pull the block at a steady rate across the surface (waxed paper, fine sandpaper, and so on). Different teams should use different surfaces. Make sure you record the surface you use. Be prepared to share the data you collect with the class. Your class should design a class data table on the board or on a transparency.

3 Put a rubber band around the block so that it is below the center (about one-fourth of the way to the top of the block as measured from the table). Attach the spring scale hook to the rubber band and pull the block so that it moves smoothly across the table, as shown in Figure 4.3. Measure the frictional force as the block slides at a constant speed along the surface. Do this for each side of the block.

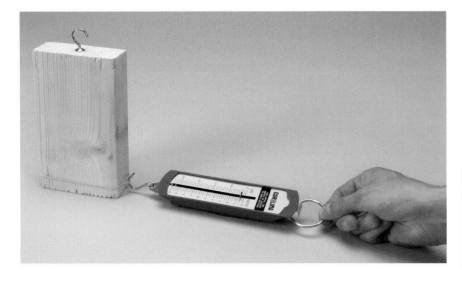

ATTACH A RUBBER BAND AND A SPRING SCALE TO THE BLOCK AS SHOWN HERE TO PULL THE BLOCK ON ONE OF ITS NARROW SIDES.
FIGURE **4.3**
PHOTO: ©2009 Carolina Biological Supply Company

4 Record your data, then discuss the results with your team and share your results with the class. Add your results to the class data table.

5 Record your answer to the following question: How did your prediction compare with the results?

6 Think about the variables. Which variables did you keep constant this time? Record your answer.

REFLECTING
ON WHAT YOU'VE DONE

1 Write answers to the following questions in your science notebook:

A. Summarize in several paragraphs what you have learned in this lesson about the force of friction. In your summary, include what you know about the factors that affect frictional force and explain how you measure it.

B. In this lesson, you measured sliding friction. Why does the force on the spring scale measure the force of friction while the block moves at a steady speed?

C. Suppose you used ice as a surface for the block to slide on. What results would you get in this lab? Consider results for all three variables— surface type, weight of the block, and surface area.

Nature Puts on the BRAKES

S kydiving is a sport that many people enjoy. But it takes courage! When skydivers leap from planes, you might think that the constant pull of gravity would make them fall faster and faster until their parachutes opened and allowed them to glide safely back to the earth. But that's not quite what happens.

At the beginning of a skydive, a diver's fall does speed up, or accelerate, rapidly. But as the diver falls, the acceleration continually decreases until the diver stops speeding up. The diver then falls at a constant velocity. ("Velocity" is the speed at which an object is traveling in a single direction—in the case of a skydiver, down!) The constant velocity that the skydiver reaches is called terminal velocity. It usually takes a skydiver about 10 seconds to reach terminal velocity.

▶ **A SKYDIVER LEAPS FROM A PLANE.**

PHOTO: U.S. Air Force photo by Staff Sgt. Matthew Hannen

Air Friction   Gravity

TERMINAL VELOCITY IS A RESULT OF THE INTERACTION OF TWO FORCES: GRAVITY AND AIR FRICTION (DRAG).

FORCES ON SKYDIVERS

It is the forces on skydivers that make them eventually reach terminal velocity. To the observer, gravity is the more obvious force. If gravity were the only force acting on skydivers, their velocity would continue to increase at a rate of 9.8 meters per second (32.2 feet per second) each second that they fall. Ten seconds into the fall, they would be moving at 98 meters (322 feet) per second if gravity were the only force on them. As they continued to fall, they would continue to go faster and faster.

But gravity isn't the only force on skydivers. Air friction, or drag, pushes up on their bodies as they fall through the air. ("Drag" is another word for the force of friction between the skydiver and the air.) The force of gravity is constant, but drag increases with the skydiver's speed. As skydivers fall, they eventually reach the point where the size of the drag force equals the size of the force of gravity. The drag continues to push up while gravity continues to pull down, and the two forces counterbalance each other. The skydivers have reached terminal velocity.

BECAUSE THE TERMINAL VELOCITY OF SPREAD-EAGLE SKYDIVERS IS VERY NEARLY THE SAME, THEY CAN FORM PATTERNS LIKE THE ONE SEEN HERE AS THEY FALL.

PHOTO: CJ Smithson/creativecommons.org

CONTROLLING TERMINAL VELOCITY

Skydivers can make their terminal velocity faster or slower by changing their body position as they fall. For skydivers falling with their backs or stomachs parallel to the earth, terminal velocity is about 50 meters (164 feet) per second after 10 seconds. If skydivers were going head first, terminal velocity would be twice that rate, or 100 meters (328 feet) per second.

Why the difference? Because terminal velocity also depends on the surface area of the diver against which the air pushes during the fall. The greater the surface area, the greater the drag. The smaller the surface area, the less the drag, and the faster the fall. To understand the comparison, think of swimmers—they feel much more resistance entering the water with a belly flop than with a nosedive.

BY CONTROLLING THE AMOUNT OF THE BODY THAT IS EXPOSED TO THE DRAG OF THE AIR, A SKYDIVER CAN CHANGE THE TERMINAL VELOCITY. HERE, A SKYDIVER FALLS IN A SPREAD-EAGLE POSITION. SINCE THE MAXIMUM AMOUNT OF BODY IS EXPOSED TO THE AIR, THE AIR FRICTION IS GREATER AND THE TERMINAL VELOCITY IS SMALLER THAN WHEN THE BODY IS POINTED STRAIGHT DOWN.

PHOTO: NASA Spinoff Magazine

▶ PARACHUTES ARE DESIGNED TO TAKE ADVANTAGE OF AIR FRICTION, ALLOWING SKYDIVERS TO LAND SAFELY.

PHOTO: U.S. Navy photo by Mass Communication Specialist 2nd Class Christopher Stephens

Falling at terminal velocity without a parachute is still too fast to land safely on the ground, so skydivers do need that parachute. Parachutes help slow skydivers even more by greatly increasing their surface area. Large parachutes are more powerful than small ones. The larger the parachute, the greater the air resistance acting on the diver, and the slower the terminal velocity. The probability of a smooth, safe landing increases proportionately! ■

DISCUSSION QUESTIONS

1. How is air friction the same or different from the sliding friction studied in this lesson?

2. What would happen to a skydiver jumping from a spacecraft near the moon?

ROCK CLIMBING:
Two People, One Powerful Force

Rock climbing is not usually thought of as a team sport. For most climbers, however, it is. The climber has a partner, called a belayer.

As the climber ascends the slope, the belayer often remains below. The job of the belayer is to keep the climber from falling. The climber or belayer places an anchor at the top of the climb or at some other convenient location. The belayer holds one end of a rope. The other end of the rope is threaded through the anchor and then attached to the climber's waist. As the climber ascends the slope, the belayer holds the rope. If the climber should slip and fall, the belayer would pull on the rope and prevent the climber from falling to the ground below.

GOING UP!

Both partners make full use of a powerful force called friction. Climbers use friction to make their way up to the top. While ascending a steep rock face, the climbers constantly look for "holds"—cracks and crevices into which they can wedge their feet, or small rocks or ledges that they can grasp or step on. If the rock face is smooth, climbers have to rely only on the friction between their hands and feet and the rock face to keep from falling.

▶ AT A TIME LIKE THIS, FRICTION COMES IN HANDY.

PHOTO: Iwona Erskine-Kellie/creativecommons.org

Climbers also increase the force of friction by wearing shoes with specially designed soles. These soles cover more of the foot than the soles of everyday shoes. This is important because this makes it easier for the climber to get some part of the sole of the shoe on the rock and create a lot of friction with the rock.

▶ BELAYING DEVICE FOR CLIMBING

PHOTO: Courtesy of Black Diamond Equipment

▶ CLIMBERS' SHOES ARE DESIGNED TO GRIP THE ROCK SURFACE AND PREVENT THEIR FEET FROM SLIPPING.

PHOTO: Cristian Ordenes/creativecommons.org

HOLDING ON!

While climbers make their way up the mountain or cliff, belayers keep an eye on their progress. Belayers use one of several methods to increase the friction on the rope and ensure that the climbers will be safe. For example, belayers can wrap the rope around their hips. The friction between their clothing and the rope wrapped around them is far greater than the friction between their hands and the ropes. Should the climbers slip, belayers can control the fall better than they could if the rope were just in their hands. More often, belayers thread the rope around special metal devices that increase the friction on the rope and decrease the force the belayer must exert if the climber falls.

Climbers and belayers aren't usually thinking about friction when they're moving up a mountainside. They're focusing on their goal: reaching the top. Friction helps keep them safer and makes the climb easier. And, unlike those special shoes and mountaineering equipment, it's absolutely free! ■

 DISCUSSION QUESTIONS

1. What are some ways in which rock climbers overcome the force of gravity?

2. What are some other situations in which gravity is counterbalanced by friction?

MAGNETIC FORCES

▶ BOATS LIKE THESE ARE USED TO PICK UP ABANDONED BIKES AND OTHER MATERIALS FROM THE CANALS IN AMSTERDAM. A CRANE WITH AN ELECTROMAGNETIC HEAD ATTACHED TO THE BOAT RETRIEVES THE UNWANTED MATERIALS FROM THE WATER.

PHOTO: reggestraat/creativecommons.org

INTRODUCTION

In Lesson 2 you investigated gravity and observed how Earth's gravity attracts objects. Another naturally occurring force is magnetism. Magnetic forces are used in many ways. Can you name ways that you have seen magnets used? In this lesson you will explore magnetic forces and conduct a "fair test" of what affects a magnet's strength.

OBJECTIVES FOR THIS LESSON

- Distinguish magnetic from non-magnetic materials.

- Observe and describe the nature of magnetic forces.

- Conduct a fair test of factors that affect a magnet's strength.

MATERIALS FOR LESSON 5

For you

1	copy of Student Sheet 5.1: What Do We Know About Magnets?
1	copy of Student Sheet 5.2: The Strengths of Different Combinations of Magnets

For your group

4	flexible magnets
1	pack of assorted materials
2	plastic cups
1	tongue depressor
1	large paper clip
25	washers

GETTING STARTED

1 On Student Sheet 5.1: What Do We Know About Magnets? record what you know and what you would like to know about magnets in Columns 1 and 2. Discuss your list with your lab partners.

2 Contribute to a class discussion about magnets. Your teacher will list all the ideas and questions you have on a sheet of newsprint.

3 Put your student sheet in a safe place so that you can complete Column 3 after you have completed Lessons 5 and 6.

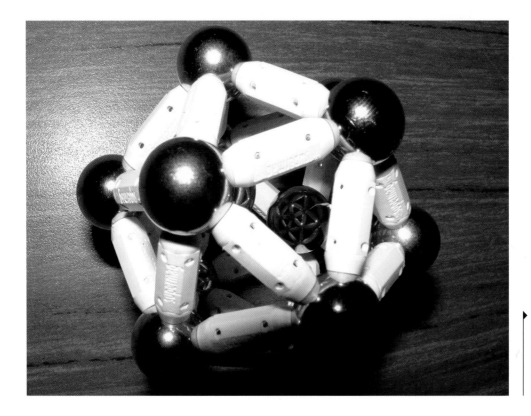

▶ **WHICH OF THE COMPONENTS IN THIS TOY DO YOU THINK ARE MAGNETIC?**

PHOTO: Juan Tello/creativecommons.org

IS IT OR ISN'T IT MAGNETIC?

PROCEDURE

1 Think about the following questions and be prepared to share your ideas with the class:

A. How can you decide whether something is magnetic or not?

B. How can you tell if something is only slightly magnetic?

C. How can you tell the difference between a magnet and something made of magnetic material?

2 Use two magnets and explore how they react when brought near each other. Test all sides of your magnets. Record your observations in your science notebook. Use your observations to answer these questions: ☞

A. What words can you use to describe the forces on the magnets when they are near each other?

B. Do the magnets have to touch to exert a force on each other?

C. What else can you conclude about the forces between magnets?

3 Can magnets exert forces on objects that are not magnets? Discuss your ideas with your lab partners.

4 Examine your pack of assorted materials. Do you think a magnet will exert forces on any of them? Before you test them to find out the answer to this question, make a data table to record each item and a prediction about whether you think a magnet will exert a force on it. What else will you include in your data table?

5 Now use a magnet to test the materials in your pack. Be sure to record your observations.

NOTE Make careful observations. The magnetic force may be very small. If you don't observe carefully, you may not detect it.

6 When you have finished, look back at your predictions. Evaluate your predictions based on the observations you made. Do you detect any patterns in your data?

7 Participate in a class discussion about what you have learned in this inquiry.

INQUIRY **5.2**

MEASURING MAGNETS

PROCEDURE

1 Do you think putting two magnets together will make them stronger or weaker than one magnet alone? What could you do to find out the answer to this question? Share your ideas with your lab partners.

2 Read about fair tests in "Playing Fair."

3 Think of different ways to test how strong a magnet is. Can you stick two magnets together to make a stronger magnet? How can you compare the strength of two magnets stuck together to one magnet using the materials given to you? Work with your lab group to develop a plan to do this with the equipment given to you.

4 Share your group's plan with the class. After listening to the class discussion, make any modifications to your plan that you feel are necessary. Figures 5.1 and 5.2 show one way of using the materials to test magnetic forces.

READING SELECTION
BUILDING YOUR UNDERSTANDING

PLAYING FAIR

When you play a game you want everyone to have an equal chance to win. To do that, you must come up with a set of rules that everyone agrees on, and everyone should follow these rules when they play the game. In science, researchers want to make sure that investigations are a "fair test" of the question that is being asked. To do this, scientists follow a set of guidelines, or rules, to ensure that an experiment is a "fair test." Scientists also call a fair test a "controlled experiment." Only one variable is changed at a time during a controlled experiment, and all other variables are kept constant. If the experiment is controlled, any changes in the dependent variable will be the result of changes in the independent variable. Fair tests or controlled experiments produce conclusions that can be validated by other scientists. So when you design an experiment, play by the rules and always make sure it is a fair test. ■

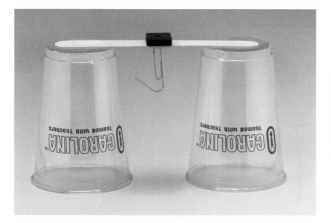

▶ **ONE WAY TO TEST THE STRENGTH OF MAGNETS**
FIGURE **5.1**
PHOTO: ©2009 Carolina Biological Supply Company

REFLECTING
ON WHAT YOU'VE DONE

1. In your notebook, write a paragraph summarizing what you have learned about magnetic forces in this lesson.

2. Suppose you taped 6 magnets together. How many washers do you think they would hold? Explain your reasoning. What about 12 magnets stuck together?

3. In Lesson 2, you investigated the force of gravity. How are gravity and magnetism alike? How are they different?

▶ **IS THIS PROCEDURE A "FAIR TEST"?**
FIGURE **5.2**
PHOTOS: ©2009 Carolina Biological Supply Company

5 Complete your experiment to find out what happens to the strength of magnets when they are stuck together. Conduct your "fair test" and record your data.

6 Discuss with your group how you can make a graph to better see the pattern in your data. Plot your graph on Student Sheet 5.2: The Strengths of Different Combinations of Magnets.

7 What can you conclude about the strength of magnets that are stuck together and the force they exert?

THE UNIQUE PROPERTY OF
LODESTONE

▶ THESE PAPER CLIPS STICK TO LODESTONE BECAUSE
IT HAS MAGNETIC PROPERTIES.

PHOTO: Chip Clark, National Museum of Natural History, Smithsonian Institution

As long ago as the sixth century B.C., an ancient Greek philosopher named Thales observed a rock with magnetic properties. It attracted objects made of iron.

About the third century B.C., legend has it that a shepherd was walking in Magnesia, a region in Asia. He noticed that his sandals seemed to be sticking to the ground. He looked about and saw lots of strange rocks. The iron nails in his sandals were clinging to these rocks! The Chinese, who also noticed these rocks, named them tzhu shih, or loving stones, because the stones liked to "kiss" other objects.

Ancient people thought the rocks were magical. Today we know the rocks were magnetic. The term magnetic is derived from Magnesia, the name of the region where that legendary shepherd walked.

It wasn't long before people found uses for the magnetic rocks. The Chinese found that if they put one of the rocks on a piece of wood and floated it on water, it aligned itself in a north-south direction. So sailors began using the rocks to guide them at sea. Then they discovered that if they rubbed needles against the rocks, the needles would also point north and south. This is how the magnetic compass was born! The rocks became known as lodestones. "Lode" means "to lead."

WHAT IS LODESTONE?

Lodestone is a pale brown to black rock. It contains lots of magnetite, a black magnetic mineral. The presence of magnetite explains in part why lodestone is magnetic. But magnetite is common in nature, while lodestones are quite rare. That's because lodestone is magnetic only when a large number of magnetic particles are aligned.

What makes some magnetite turn into lodestones? One possible explanation is lightning. Lightning carries an electric current that creates a strong magnetic field. When magnetite is exposed to a strong magnetic field, it becomes a permanent magnet. So, if lightning strikes one of these magnetite-rich rocks—Zap!—the rock turns into a strong magnet. A lodestone is created.

THEY'RE "MAGIC"!

For centuries, people believed lodestones had magical powers. Magicians used them to tell fortunes. Ancient healers thought lodestones could cure just about anything, from injuries, cramps, and asthma to poor eyesight and a weak heart.

▶ GET A "LODE" OF THIS – LIGHTNING CREATES LODESTONES!

PHOTO: NOAA Photo Library, NOAA Central Library, OAR/ERL/National Severe Storms Laboratory (NSSL)

Others thought lodestones were a kind of spirit. Some regularly dunked the stones in water, then drank the water. They believed that the water had magical powers that could protect them. Many people believed lodestones would ward off evil, and carried them as good-luck charms.

Although we now understand how lodestones work, their magnetic force still inspires wonder. ∎

DISCUSSION QUESTIONS

1. In what ways has the discovery of magnetite changed daily life?

2. What other scientific words or element names (besides magnet) can you think of that come from common names or locations?

MAGNETS MADE FROM DIFFERENT MATERIALS

Although magnets can be made of different materials, most are made from steel, alnico, ferrite, and rare-earth metals.

ALNICO MAGNETS

These magnets get their name from three of the materials they're made from: Al (aluminum), Ni (nickel), and Co (cobalt). Strong, tough, and heat-resistant, alnico magnets can be found in motors, microphones, and loudspeakers.

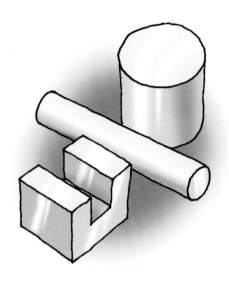

MANY ALNICO MAGNETS ARE SPECIAL SHAPES AND ACT AS STRONG MAGNETS IN DIFFERENT DEVICES.

▶ FERRITE MAGNETS COME IN ALL SHAPES AND SIZES.

FERRITE MAGNETS

"Ferrite" comes from the Latin word for iron. To make ferrite magnets, manufacturers blend iron with other materials. Ferrite blended with plastic makes soft, flexible magnets. Ferrite blended with clay can be formed into shapes and then baked like pottery to make ceramic magnets.

Ferrite magnets are all around you. You may use them to attach papers and photos to your refrigerator. These magnets are also found in toys, small motors, and door latches.

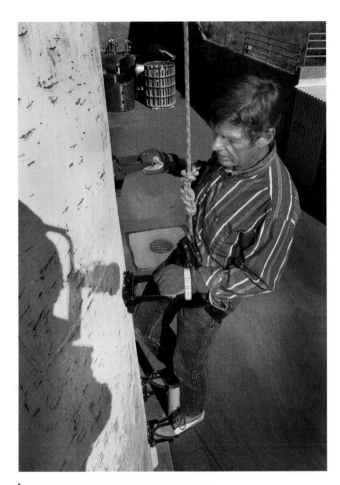

THIS MAN IS USING MAGNETS TO CLIMB A STEEL WALL.

PHOTO: Courtesy of Los Alamos National Laboratory

RARE-EARTH MAGNETS

Rare-earth magnets are made of rare metals. Two examples are neodymium and samarium. Rare-earth magnets are extremely powerful. If you strapped large ones to your hands and feet, you could climb a steel wall, as the man at left is doing. Devices that require small, strong magnets—computer disk drives, satellite systems, and headphones—use rare-earth magnets. ■

DISCUSSION QUESTIONS

1. Why don't we make all of our magnets out of the strongest materials?

2. Invent some futuristic uses for magnets and decide whether you would use alnico, ferrite, or rare-earth magnets.

LESSON 6

THE EARTH'S MAGNETIC FORCE

► **COMPASSES COME IN HANDY WHEN YOU NEED TO FIND YOUR WAY AROUND.**

PHOTO: © Roy McMahon/Corbis

INTRODUCTION

In Lesson 5 you investigated magnetic forces. You saw how magnets can exert forces on other magnets and on other objects that can be magnetized. In this lesson you will investigate the magnetic force that is the result of the earth's magnetic field at your location. You will build a compass and use it to see how the force of the earth's magnetic field affects the compass.

OBJECTIVES FOR THIS LESSON

Discover the characteristics of a magnetic compass.

Develop proficiency in manipulating materials, following plans, and troubleshooting.

Investigate magnetic poles.

Learn about the historical significance of the magnetic compass and its present uses.

▶ MATERIALS FOR LESSON 6

For you

Your copy of Student Sheet 5.1: What Do We Know About Magnets?

For your group

3	flexible magnets
1	plastic drinking straw
2	pieces coated hook-up wire, 20 cm in length
1	straight pin
1	plastic cup
2	colored stickers
1	magnetic compass

GETTING STARTED

1 Discuss the following questions with your classmates:

    A. How are maps useful to us?

    B. What kind of information can you get from a map?

    C. If you needed to find out which way north is, how would you do it?

2 Your teacher will now give you a magnetic compass to work with. Take turns with your partner manipulating it. Be sure to hold it flat. As you explore the compass, discuss the following questions with your partner:

    A. Why do you think most compasses point the same way?

    B. How can you make the needle point in different directions?

    C. How could you test to see if the needle (pointer) is a magnet?

3 Read "Wayfinding" on pages 64-66, which describes different compasses. Which compasses did you like best?

INQUIRY 6.1

BUILDING YOUR OWN COMPASS

PROCEDURE

1 Pick up the materials you will need to build your compass (see Figure 6.1). Follow the directions on the next three pages to build your compass.

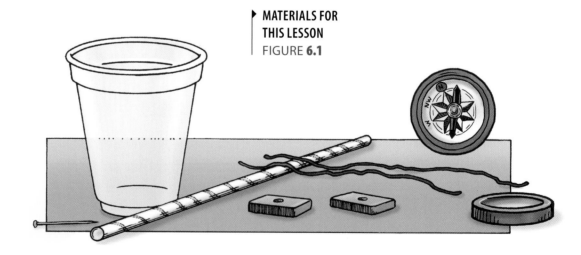

▶ **MATERIALS FOR THIS LESSON**
FIGURE **6.1**

2 Bend the two wires in half around your finger as in Figure 6.2. You are doing this so that the wire will fit snugly inside the straw when you get to Step 5.

3 Arrange the two magnets so that they stick together side by side, as shown in Figure 6.3. Push the open ends of one wire one third of the way through the hole in the center of one of the magnets. Now do the same for the other magnet with the other wire.

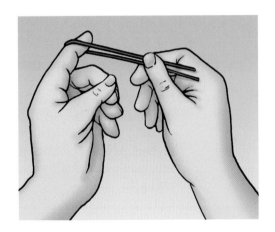

▸ BEND THE WIRES
AROUND YOUR
FINGERS THIS WAY.
FIGURE **6.2**

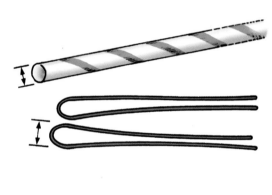

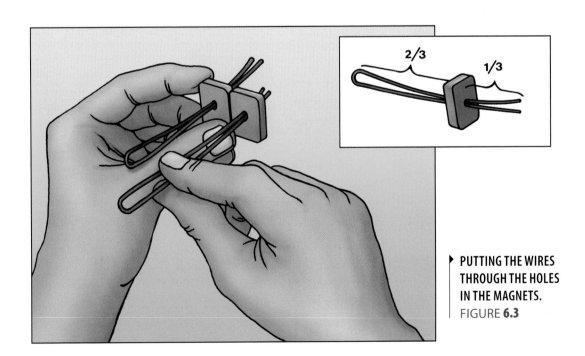

▸ PUTTING THE WIRES
THROUGH THE HOLES
IN THE MAGNETS.
FIGURE **6.3**

Inquiry 6.1 continued

4 Without removing the wire, separate the two magnets. Pull apart the open ends of the wire and wrap each end around the long sides of the magnet. Bend the looped end up over the top and let it project forward as shown on Figure 6.4. Now do the same to the other magnet.

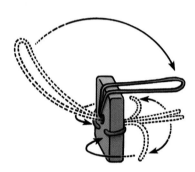

▸ **BEND THE WIRE AS SHOWN HERE.**
FIGURE **6.4**

5 Check to see if the sides of the magnets away from the loop will attract each other. If they repel, go back to Step 2, remove the wire from one of the magnets, turn it over, and rewire it. The magnets should stick together with the looped ends of the wire pointing in opposite directions, as in Figure 6.5.

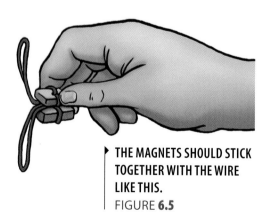

▸ **THE MAGNETS SHOULD STICK TOGETHER WITH THE WIRE LIKE THIS.**
FIGURE **6.5**

6 Insert the looped ends into the ends of the straw. See Figure 6.6.

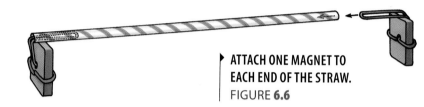

▸ **ATTACH ONE MAGNET TO EACH END OF THE STRAW.**
FIGURE **6.6**

7 Balance the straw on your finger (see Figure 6.7) and push the pin down through the straw at the place where it balances. The tip of the pin should project about 5 mm below the bottom of the straw. Do not make more than one or two holes in the straw or the pin will slip.

▸ **BALANCE THE STRAW WITH THE MAGNETS ON YOUR FINGER. PUT A PIN IN THE STRAW AT THE BALANCE POINT.**
FIGURE **6.7**

8 Set the straw with the pin in it so that the pin is near the middle of the inverted cup as shown in Figure 6.8.

NOTE Do not push the pin into the plastic cup.

Figures 6.9 and 6.10 show two ways to adjust the balance of your compass.

This kind of adjusting is called troubleshooting. Be patient and adjust one thing at a time. You will get your compass to balance. If you are having difficulty, keep trying. You will discover how to make your compass work.

9 Do not disassemble your compass. Store your compass as directed by your teacher. You will use your compass again in the next inquiry.

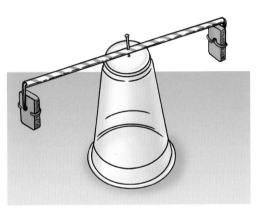

▶ SET THE STRAW WITH THE PIN IN IT ON THE TOP OF THE INVERTED CUP. YOU WILL PROBABLY NEED TO ADJUST THE BALANCE.
FIGURE **6.8**

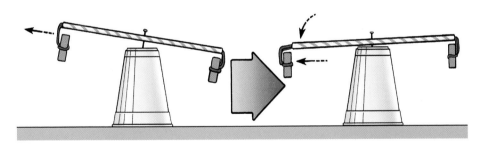

▶ YOU CAN MOVE THE MAGNETS IN OR OUT OF THE END OF THE STRAW TO ADJUST THE BALANCE.
FIGURE **6.9**

▶ YOU CAN TWIST OR ROTATE THE MAGNETS A BIT TO ADJUST THE BALANCE.
FIGURE **6.10**

USING YOUR COMPASS: WHICH WAY IS WHICH?

PROCEDURE

1 Get your compass and make sure it is balanced. You may have to make a few adjustments if the compass was bumped.

2 After you have balanced your compass, rotate the straw to different positions and release it. Make sure you try several different release positions. Record what happens in your science notebook. ✐

3 Look at the other compasses in the room and observe which way they are pointing after they have been released and have "settled down."

4 When your teacher asks you, mark the north side, or "pole," of your compass with one of the colored stickers.

5 Now work with another lab group to investigate how your compass behaves. Follow the steps described below and record your observations and responses:

A. Begin by moving the compasses very slowly toward each other. What happens? Do you think opposites attract?

B. Now that you know which poles attract and repel each other, try to decide which side of your extra magnet (the unmarked one not used in the compass) is the north pole. Discuss your method with your lab group.

C. Mark the side that your group thinks is the north pole of the magnet with the second colored sticker. Check your ideas by seeing how one of the compasses moves when you bring the north pole of the magnet near.

6 Describe how you determined which side of the extra magnet is the north pole.

7 Use the extra magnet to make the compass spin. Take turns and try to find as many different ways as you can. Describe the different ways that you were able to make the compass spin.

1 In your science notebook, record your responses to the following:

A. Identify the forces that you observed in this inquiry. What were you able to do with these forces?

B. Based on your observations in this inquiry, what can you conclude about the earth?

C. Look at the world map on the wall. Suppose you could take your compass to different places around the world. In what direction do you think your compass would point at different locations around the world?

D. Write a definition of a compass based on what you have learned in this lesson.

2 Review your Student Sheet 5.1: What Do We Know About Magnets? Complete the column "What I Have Learned."

3 Read "Do Animals Use Magnetism?" Write a response to what you read in your notebook.

EARTH'S MAGNETIC PERSONALITY

▶ **LAVA ERUPTS FROM MOUNT KILAUEA**

PHOTO: H. Powers/U.S. Geological Survey

Mount Kilauea, a volcano on the island of Hawaii, erupts. Red-hot lava oozes down the mountain. Once the lava cools, scientists can chip glassy rocks from the hardened layers of lava. They can learn a lot about Earth's history by studying these rocks.

Richard Fiske, a geologist with the Smithsonian Institution's National Museum of Natural History in Washington, D.C., has brought a variety of lava rock samples back from Hawaii. One reason he's interested in lava rocks is that they contain tiny pieces of a mineral called magnetite. This is the same mineral found in lodestones. Just what is so special about the magnetite in lava rocks?

READ IT IN THE ROCKS

At low temperatures, magnetite behaves like a small magnet, but if the temperature gets too high, magnetite loses its magnetic properties. So any magnetite present in the lava pouring out of an erupting volcano loses its magnetism.

Now here's the interesting part. As lava cools, its magnetism returns because of the influence of the magnetic field surrounding the lava. A magnetic field is a way of describing the magnetic force produced by a magnet or current.

What magnet is causing the magnetic field? It's Earth, which acts as a giant magnet. As the magnetite cools, it becomes magnetized in the direction of the Earth's magnetic field.

Scientists use an instrument called a magnetometer to measure the direction and intensity of the lava rocks' magnetism. They also use other techniques to determine the age of the rocks. With these measures, scientists can tell exactly in which direction Earth's magnetic field pointed at the time that rock formed. Each time a volcano erupts, a new layer of rock forms on top of the old one. The layers record the movement of Earth's magnetic poles over time.

▶ **A GEOLOGIST COLLECTS LAVA ROCKS, WHICH CONTAIN A MINERAL CALLED MAGNETITE.**

PHOTO: Richard S. Fiske, National Museum of Natural History, Smithsonian Institution

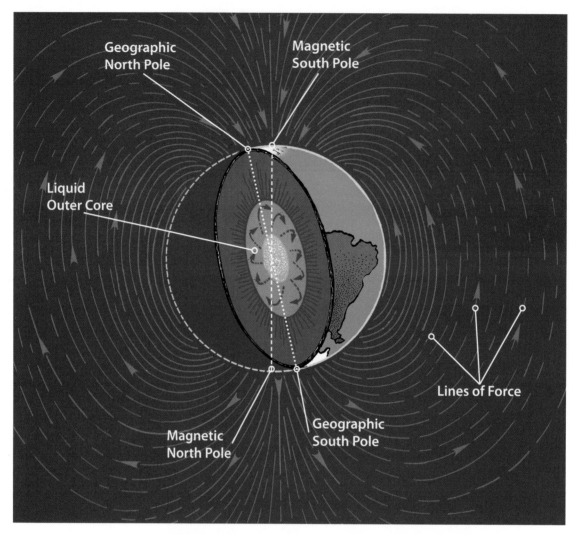

Geographic
North Pole

Magnetic
South Pole

Liquid
Outer Core

Magnetic
North Pole

Geographic
South Pole

Lines of Force

▶ THIS IS AN ILLUSTRATION OF EARTH'S MAGNETIC FIELD. NOTICE THE TWO SETS OF POLES: MAGNETIC AND GEOGRAPHIC. THE GEOGRAPHIC NORTH AND SOUTH INDICATE EARTH'S POINT OF ROTATION. THE MAGNETIC POLES ARE WHERE EARTH'S MAGNETIC FORCE IS GREATEST. EARTH'S MAGNETIC POLES ARE ABOUT 11 DEGREES FROM EARTH'S POINT OF ROTATION. THE NORTH MAGNETIC POLE IS IN THE SOUTHERN HEMISPHERE WHILE THE SOUTH MAGNETIC POLE IS IN THE NORTHERN HEMISPHERE.

ROCKS AND REVERSALS

These records reveal that the position of Earth's magnetic poles changes from time to time. This discovery was first made in the 1950s. Scientists studying a series of old lava flows stumbled upon something surprising. In one lava layer, the magnetic direction of the magnetite was just as expected—north was north and south was south. But in the very next layer the magnetic direction of the particles flip-flopped, and north became south!

"At first, people were stunned when they saw these reversals," says William Melson, another scientist at the Smithsonian's National Museum of Natural History. "They thought perhaps Earth itself had once flipped upside down. But then they realized that was impossible. We began to understand that it was Earth's magnetic field that had reversed itself."

CAUSE OF THE REVERSALS

During the last 75 million years, there have been numerous reversals of the magnetic field. Why do the magnetic poles shift? No one is completely sure, but scientists think that it has something to do with the interaction between hot, liquid iron in Earth's outer core and Earth's rotation, which creates electric currents. These currents in turn produce large and complex magnetic fields.

Scientists speculate that a change in circulation patterns in the outer core causes the magnetic poles to reverse.

The last reversal was approximately 700,000 years ago. Magnetic south moved from Antarctica back to where it is currently found—in the Arctic. Magnetic north went back to Antarctica. Scientists think that the magnetic poles will reverse again, although they can't predict exactly when that will be. ∎

DISCUSSION QUESTIONS

1. What is the difference between Earth's geographic poles, which are commonly called the North and South poles, and Earth's magnetic poles?

2. Are Earth's magnetic poles always in the same place? Discuss the forces that act upon them.

LODESTONE SPOON COMPASS
PHOTO: © Exploratorium, www.exploratorium.edu

No, this is not a soup ladle. It's one of the first compasses, a lodestone carved to look like a spoon. Chinese inventors made it in about 100 A.D. When the spoon-shaped lodestone was placed on a brass plate, the spoon rotated until its handle pointed south.

From this primitive example to today's sophisticated devices, people have used many variations of the magnetic compass. Take a look at how the compass has changed through the years.

EARLY COMPASSES

A MAGNETIZED NEEDLE

During the 11th century, people in China made a compass by using wax to attach a small iron needle to a strand of silk. They rubbed the needle with a lodestone. This permanently magnetized the needle. They discovered that the magnetized needle pointed north and south. The end of the needle that pointed north came to be known as the "north pole" of the magnet. Since opposite poles attract, the north pole of the compass needle points to the magnetic south pole, which is near the north geographic pole.

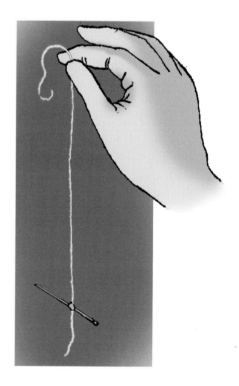

▶ **AN EARLY COMPASS**

FLOATING COMPASSES

The Chinese invented another compass in the 12th century. It was a floating compass. People made these compasses by sticking a magnetized iron needle crosswise through a small piece of straw and floating it in a bowl of water. Then they moved a lodestone around the bowl, causing the needle to spin. Next, they moved the stone away from the compass. The needle continued to spin until it finally settled, pointing north. The floating compass was used for more than 500 years.

MOVING THE LODESTONE AROUND THE BOWL CAUSES THE MAGNETIZED NEEDLE TO SPIN.

EARLY COMPASS CARDS HAD EIGHT POINTS, MATCHING THE DIRECTIONS OF THE WIND THAT SAILORS WERE FAMILIAR WITH AT SEA.

PIVOTED COMPASSES

In a pivoted compass, which was used during the 1300s, the magnetic needle is balanced on a pivot, or a small pin. It wasn't long before the needle and pivot were attached to a compass card.

THE NEXT GENERATION OF COMPASSES

By the 1500s, the compass had started to look like the instrument we use today. The magnetic needle was attached on a pivot to a circular card with 32 points, or directions. Christopher Columbus used this kind of compass.

In 1745, the Englishman Gowin Knight developed a steel needle that kept its magnetism much longer than an iron needle did. He built a compass with the needle. It became known as the Knight compass.

THE MARINER'S COMPASS

On board ships, compasses were suspended from a string. This design was used so that the motion of the ship wouldn't affect them. ■

▶ **THE CARD IN THIS COMPASS, WHICH WAS USED IN THE LATE 1500S, HAS 32 POINTS.**

PHOTO: © National Maritime Museum, Greenwich, London

▶ **THIS IS AN EXAMPLE OF A MARINER'S COMPASS.**

PHOTO: National Science Resources Center

DISCUSSION QUESTIONS

1. What are the benefits and drawbacks to the different kinds of compasses that have been invented?

2. When might you have an occasion to use a compass today?

Do Animals Use Magnetism?

SOME STUDIES INDICATE THAT PIGEONS USE MAGNETISM TO FIND THEIR WAY HOME.

PHOTO: mill56/creativecommons.org

If you were lost in the middle of the woods and couldn't see the sun, you might use a compass to try to decide which direction to take. A magnetic compass needle lines itself up with Earth's magnetic field and points roughly north and south: from that, you can figure out east and west, too. Because this works fairly well, people have been using magnetic compasses to find their way for about 1,000 years.

But how do other animals find their way? How do they navigate when it is cloudy? You probably know that many animals rely on their sense of smell to keep track of where they have been and where other animals are. However, some animals migrate (travel from one place to another), regularly covering hundreds or even thousands of kilometers each year. It seems unlikely that animals could repeat such long trips accurately if they were relying only on their sense of smell, so scientists have been looking for evidence of what else animals may use to navigate. There are scientific investigations of whether animals use the sun and moon, Earth's magnetic field, and recognition of landmarks to repeat their long journeys.

Homing pigeons are famous for being able to navigate over extremely long distances. Their 'homing' is so reliable that they were used in World War I and World War II to deliver messages over enemy lines. How do homing pigeons find their way—even on cloudy days? Do they carry a map and a compass?

Researchers have discovered a small spot on the beak of pigeons and some other birds that contains magnetite. Magnetite is a magnetized rock, which may act as a tiny GPS unit for the homing pigeon by giving it information about its position relative to Earth's poles. Researchers have also found some specialized cells in birds' eyes that may help them see magnetic fields. It is thought that birds can use both the beak magnetite and the eye sensors to travel long distances over areas that don't have many landmarks, such as the ocean.

Other animals besides birds make use of electromagnetic fields. Sharks have jelly-

▶ SCIENTISTS ARE CURRENTLY STUDYING HOW SEA TURTLES USE MAGNETISM TO HELP THEM NAVIGATE.

PHOTO: Leonard Low/creativecommons.org

filled canals on their heads that are extremely sensitive detectors of electric fields in the water. It is thought that they use these "ampullae" to find prey (the muscles of the prey make electric fields when they contract). Sharks may also use them to find their way around by detecting the weak electric fields produced by ocean currents moving in the magnetic field of the earth. A shark repellant has been developed for scuba divers that sends out strong electric signals, which irritate the sensitive ampullae of great white sharks.

Some species of sea turtles, including leatherbacks and loggerheads, appear to use Earth's magnetism to navigate from nearly the moment they are born. First the hatchlings use reflected moonlight on the water to travel from their nests to the ocean. But when they reach the ocean, they have to follow migration routes that eventually, after hundreds of kilometers, bring them back to their place of birth. Researchers have found that hatchlings use the strength and direction of Earth's magnetic field to do these long migrations. If a small magnet is attached to their backs, they head off in the wrong direction.

Even some bacteria have magnetite in their cells that allows them to migrate along Earth's magnetic field lines. When they swim through water near a strong magnet, they all swim in one direction—depending on where the magnet is placed. In humans, deposits of magnetite have been found in bones in our noses. Do you think that we use Earth's magnetic field to know which way we are headed? ■

DISCUSSION QUESTIONS

1. What sorts of experiments might scientists perform to study how animals navigate?

2. What devices do humans use to navigate?

ROLLING ALONG

▶ **HAVING FUN WITH MOTION.**

PHOTO: Rich Moffitt/
creativecommons.org

INTRODUCTION

What happens when an object moves? How can you measure its motion? What forces are at work? These are questions that you will explore in this lesson and the lessons that follow. Understanding motion and forces makes it possible to predict where an object will be at some future time. This makes it possible to make planes fly, cars move, and even to send spacecraft to distant planets and keep them on course over billions of miles. In this lesson you will begin your investigation of motion by looking at how a steel ball moves. You will also begin to think about energy associated with moving objects.

OBJECTIVES FOR THIS LESSON

Observe and describe the motion of an object.

Learn how to calculate average speed.

Measure the speed of an object across a flat surface.

Explore Newton's First Law of Motion.

Calculate the kinetic energy of a moving object.

▶ MATERIALS FOR LESSON 7

For your group

1	steel ball
8	student timers
1	meterstick
	Masking tape

GETTING STARTED

1 Read "Measuring Motion." What does the term "average speed" mean? How do you calculate it? Discuss these questions with the class.

2 Complete the following two exercises to practice calculating average speed.

A. A car travels 100 meters (m) in 2 seconds. What is its average speed?

B. A jogger runs 50 meters in 25 seconds. What is the jogger's average speed?

3 Your teacher will give you a steel ball. What do you think the motion of the ball will be like if it is rolled across the floor? Discuss your ideas with your lab partners. Be prepared to share your ideas with the class.

▶ HOW COULD YOU DETERMINE HOW FAST THIS CAR IS GOING?

PHOTO: National Science Resources Center

MEASURING MOTION

We see motion everywhere. Think, for instance, of joggers, runners, swimmers, cars, and buses. Motion is easy to recognize. It can be described with words such as "fast" and "slow." But these words do not describe motion as precisely as scientists like to describe it. How fast is fast? How slow is slow? Fast to one person may seem slow to another. To help deal with these differences, scientists have developed a way to describe an object's motion. They measure or calculate the speed of objects that are moving. When they do that, they can easily compare the motion of fast and slow objects.

What is speed? When something is moving, it is changing its position. It takes time for this change to happen. Speed tells how fast the object changes its position. How do you measure or calculate the speed with which this happens? The speed of an object is calculated by dividing the distance it traveled by the time of travel.

The speedometer on a car measures the speed of the car. It tells how far the car travels during a given time period. For example, if the speedometer registers 96 kilometers per hour (60 miles per hour), then the car is changing its position by 96 kilometers every hour. You will travel 96 kilometers each hour you ride in the car.

Is 96 kilometers per hour fast? That depends on how the speed compares with the speed of other things. For example, a jogger may have a speed of 5 kilometers per hour. Compared with the jogger, the car is moving fast. But how does the speed of the car compare with that of a plane moving 800 kilometers per hour? That comparison makes the car seem to be moving pretty slowly.

By using speed to measure the motion of things, we can compare motion in a meaningful way. If the speed of an object varies during a time interval the way the speed of a car does, the total distance traveled divided by the time to travel the distance is called the average speed.

How do you know the speed of objects if they don't have speedometers? You need two measurements. One measurement is how far an object has traveled. The other measurement is how much time it takes to travel that distance. The rate at which the object is changing its position—its average speed—can be calculated using the following equation:

$$\text{SPEED} = \frac{\text{DISTANCE TRAVELED}}{\text{TIME OF TRAVEL}}$$

For example, if a car travels 200 kilometers in 4 hours, its speed is 200 kilometers divided by 4 hours, which equals 50 kilometers per hour.

In this lesson, you will use this definition of speed to calculate the speed of a rolling ball. To calculate accurate values for the speed, you will measure the distance traveled with a meterstick. You also need to measure the time it takes the ball to travel each distance with a stopwatch. ■

INQUIRY **7.1**

THE MOTION OF A STEEL BALL

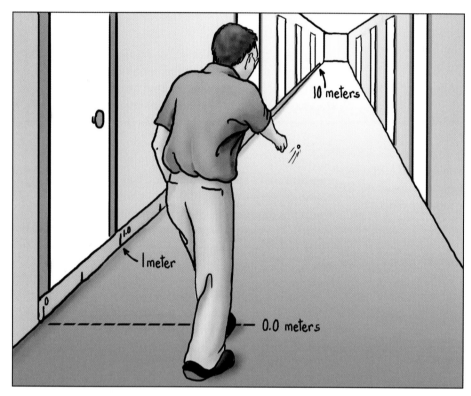

10 meters

1.0

1 meter

0.0 meters

▶ ROLLING THE
STEEL BALL
FIGURE **7.1**

PROCEDURE

1 Watch as your teacher rolls the steel ball across the floor (see Figure 7.1). Carefully observe the motion. Your teacher may need to do this more than once.

2 Discuss the following questions with your lab partner, then write your answers in your science notebook: 📝

A. Is the motion of the ball nearly constant or is it constantly changing as it moves across the floor?

B. How could you provide evidence to support your answer?

C. Why do you think the ball moves the way it does across the floor?

3 Work with your class and the materials provided to design a procedure to measure the speed of the ball as it moves across the floor. Write down the procedure the class decides on. Be sure to include a data table.

4 Work with your lab group to carry out the procedure and record data about the motion of the ball.

5 What can you conclude from your data? Record your conclusion.

6 Propose an explanation for the motion of the ball after you release it and let it roll across the floor.

BUILDING YOUR UNDERSTANDING

ENERGY OF MOTION

Energy is associated with moving objects. It is called energy of motion, or kinetic energy. Any time an object moves, it has kinetic energy. The kinetic energy of an object depends on how fast it is moving and how much mass it has. The greater the speed of an object, the greater its kinetic energy. The greater the mass of an object, the greater its kinetic energy. It is possible to calculate the kinetic energy of moving objects. You can do this by using the following formula for kinetic energy:

KINETIC ENERGY = ½ MASS TIMES SPEED SQUARED

OR

$KE = \frac{1}{2}mv^2$

If the mass is measured in kilograms (kg) and the speed is measured in meters per second (m/s), then the kinetic energy will be measured in joules (J). Since one form of energy can be converted into other forms of energy, it is useful to measure all forms of energy in joules. ■

INQUIRY 7.2

THINKING ABOUT ENERGY

1 Discuss the following question with your lab partners: Can you associate any energy with the moving ball? If so, what kind of energy do you think it is?

2 Read "Energy of Motion" on page 75.

3 Your teacher will tell you the mass of the steel ball. You can use the mass and your estimation of the speed of the ball to calculate its kinetic energy (energy of motion) as it was moving across the floor.

NOTE Be sure to record the mass in kilograms for this calculation so your kinetic energy will have units of joules (J).

4 Record your calculation in your science notebook.

5 What would happen to the speed of the ball if you were to give it a harder push? What would happen to the kinetic energy if you were to give the ball a harder push? If you have time, try this and record your calculations.

REFLECTING ON WHAT YOU'VE DONE

1 Discuss the following with your lab partner, then record your ideas in your science notebook:

A. Why did the ball move the way it did across the floor?

B. Was a force needed to keep the ball moving after it was released?

C. How did the ball get its kinetic energy?

D. Did the kinetic energy of the ball change significantly as it moved across the floor?

E. Based on what you have done in this lesson, what is the motion of a moving object if no force acts on it? What makes the object move this way?

COASTING TO PLUTO

▶ **THIS IS AN ARTIST'S RENDERING OF THE *NEW HORIZONS* SPACECRAFT APPROACHING PLUTO IN THE YEAR 2015.**

PHOTO: NASA/Johns Hopkins University Applied Physics Laboratory/Southwest Research Institute

In 2006, NASA launched a space probe named *New Horizons* on an exciting mission. *New Horizons* is headed to one of the most distant objects in our solar system, Pluto, 5 billion kilometers (3 billion miles) away. To get there, the probe has to travel very fast for a very long time. How can the probe make such a long journey? Will it need a lot of fuel to travel so fast and so far?

READING SELECTION

EXTENDING YOUR KNOWLEDGE

The probe will not need a lot of fuel to make its long journey. That is because of a very important law of motion, the law of inertia. Galileo discovered the law of inertia in the early 17th century. Newton refined Galileo's law and included it in his famous three laws of motion. In fact, he made it the First Law of Motion.

Just what does the law of inertia say? It says that an object at rest will remain at rest, and an object in motion will move at a constant speed and in a straight line if no unbalanced forces act on it. In other words, the natural tendency of an object is to continue its motion unless forces act on it to change the motion. Inertia is the property of matter that makes an object tend to maintain constant motion.

How does that apply to our speedy probe?

▶ THIS GRAPHIC DEPICTS THE PROJECTED PATH OF THE *NEW HORIZONS* SPACECRAFT THROUGH THE SOLAR SYSTEM. (NOT TO SCALE.)

PHOTO: NASA/Johns Hopkins University Applied Physics Laboratory/Southwest Research Institute

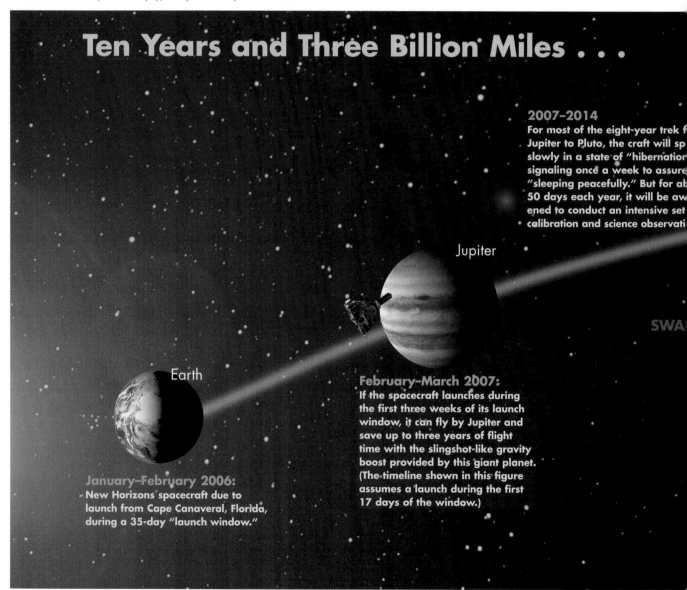

Ten Years and Three Billion Miles . . .

2007–2014
For most of the eight-year trek f
Jupiter to Pluto, the craft will sp
slowly in a state of "hibernatior
signaling once a week to assure
"sleeping peacefully." But for at
50 days each year, it will be aw
ened to conduct an intensive set
calibration and science observati

Jupiter

SWA

Earth

February–March 2007:
If the spacecraft launches during the first three weeks of its launch window, it can fly by Jupiter and save up to three years of flight time with the slingshot-like gravity boost provided by this giant planet. (The timeline shown in this figure assumes a launch during the first 17 days of the window.)

January–February 2006:
New Horizons spacecraft due to launch from Cape Canaveral, Florida, during a 35-day "launch window."

A MATTER OF FORCES AND ENERGY

New Horizons started its mission at rest relative to Earth. But it was sitting atop a powerful rocket. When the rocket was launched, the engines produced a very large unbalanced force that made the rocket with the probe on it gain a lot of speed and kinetic energy. When the right speed was reached (about 57,000 km/h or 35,000 mph), the rocket released the probe and sent *New Horizons* on its way to Pluto. The force of the rocket no longer acted on *New Horizons*, but because of its inertia, *New Horizons* kept moving. In fact, *New Horizons* is the fastest moving probe ever launched.

But there are forces that act on the probe during its long journey. As *New Horizons* travels to Pluto the Sun's gravity pulls on the probe and steadily converts some of its kinetic energy into gravitational potential energy. This conversion of kinetic energy to potential energy slows *New Horizons* down. However, because it is moving so fast, *New Horizons* has lots of kinetic energy.

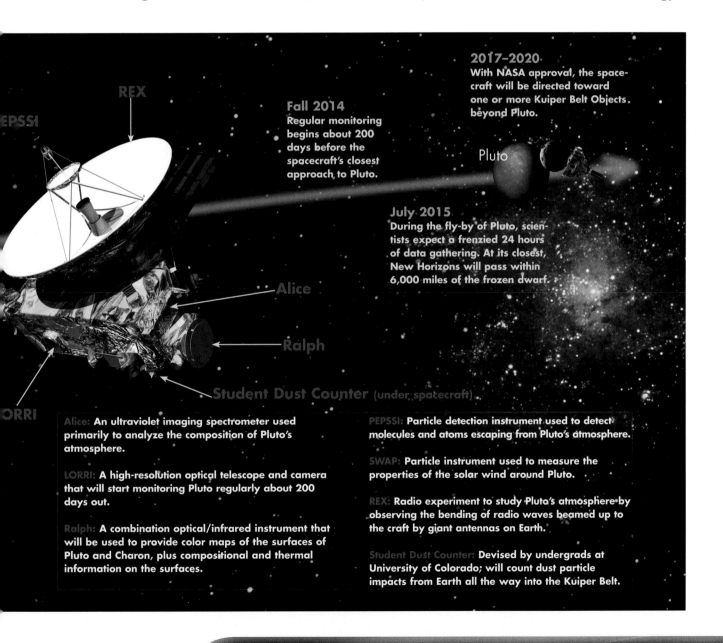

REX

EPSSI

Fall 2014
Regular monitoring begins about 200 days before the spacecraft's closest approach to Pluto.

2017–2020
With NASA approval, the spacecraft will be directed toward one or more Kuiper Belt Objects beyond Pluto.

Pluto

July 2015
During the fly-by of Pluto, scientists expect a frenzied 24 hours of data gathering. At its closest, New Horizons will pass within 6,000 miles of the frozen dwarf.

Alice

Ralph

Student Dust Counter (under spatecraft)

LORRI

Alice: An ultraviolet imaging spectrometer used primarily to analyze the composition of Pluto's atmosphere.

LORRI: A high-resolution optical telescope and camera that will start monitoring Pluto regularly about 200 days out.

Ralph: A combination optical/infrared instrument that will be used to provide color maps of the surfaces of Pluto and Charon, plus compositional and thermal information on the surfaces.

PEPSSI: Particle detection instrument used to detect molecules and atoms escaping from Pluto's atmosphere.

SWAP: Particle instrument used to measure the properties of the solar wind around Pluto.

REX: Radio experiment to study Pluto's atmosphere by observing the bending of radio waves beamed up to the craft by giant antennas on Earth.

Student Dust Counter: Devised by undergrads at University of Colorado; will count dust particle impacts from Earth all the way into the Kuiper Belt.

READING SELECTION
EXTENDING YOUR KNOWLEDGE

SOME HELP ALONG THE WAY

Fortunately, *New Horizons* also gets some help along the way. When *New Horizons* passes near a planet, the force of the planet's gravity pulls on the probe and changes the direction it is going. If the spacecraft speeds by the planet in the same general direction as the planet's motion, the pull of the planet's gravity increases *New Horizons'* speed and kinetic energy. The planet loses exactly the same amount of energy that the probe gains, but because planets are so large, the small amount of missing energy does not affect them. This is called a "gravity assist."

NASA engineers used a gravity assist when *New Horizons* passed very close to Jupiter in February 2007. The gravity of Jupiter sped up the probe, increased its kinetic energy, and helped it on toward Pluto.

When *New Horizons* left Jupiter its speed was 77,000 km/h (47,000 mph), which is very fast indeed. Even leaving Jupiter at that speed, *New Horizons* will not reach Pluto until 2015. But stay tuned, there should be some very interesting news to report from the outer solar system then. ∎

▶ PLUTO IS VERY FAR AWAY. EVEN IN LARGE TELESCOPES, PLUTO ONLY APPEARS AS A DOT AMONG STARS, AS INDICATED BY THE WHITE ARROW.

PHOTO: NASA/Johns Hopkins University Applied Physics Laboratory/Southwest Research Institute

DISCUSSION QUESTIONS

1. What forces act on *New Horizons* as it travels through the solar system?

2. If you were on the mission control team for *New Horizons*, what concerns would you have about whether and when the probe will make it to Pluto?

A hockey player cocks his arm, bringing up a hockey stick. He brings his arm down with explosive force, launching the hockey puck at a speed that can exceed 160 kilometers per hour (100 miles per hour). Hockey players do that all the time. The puck crosses the rink hundreds of times during a hockey game. What's unusual is that as it travels from one end of the rink to the other, the puck doesn't seem to slow down or change direction until an unbalanced force, like an opponent's hockey stick, acts on the puck. It seems to continue on its way to the other end of the rink at a constant speed and direction. Magic? No: inertia.

Ice hockey is an example of Newton's First Law of Motion—the law of inertia—in action. The icy surface of the rink reduces the forces that normally would slow the puck down—friction and air resistance. You know from your inquiries that smoother surfaces exert less friction. Smooth ice exerts less friction than just about any other substance, and rink ice is carefully maintained to make it as smooth as possible. Air resistance against the puck is minimal.

Since these two forces—friction and air resistance—are so small, they do not significantly affect the puck's motion. Once in

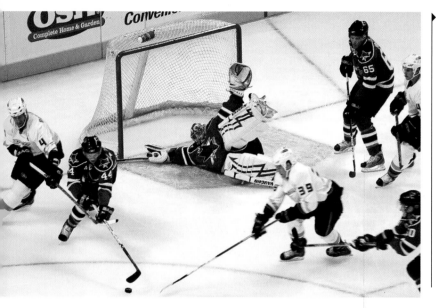

▶ A HOCKEY PLAYER'S INERTIA IS GREATER THAN A PUCK'S INERTIA. FOR ONE THING, A HOCKEY PLAYER HAS A LOT MORE MASS THAN A PUCK. BECAUSE OF THIS MASS, IT TAKES A HUGE AMOUNT OF LEGWORK FOR A HOCKEY PLAYER TO CHANGE SPEED AND DIRECTION. SOMETIMES THEY CHANGE SPEED OR DIRECTION WITHOUT MEANING TO, COURTESY OF A SHOVE FROM AN OPPONENT. A PLAYER PRESSES INTO THE ICE HARDER THAN A PUCK DOES BECAUSE OF THEIR MASS, WHICH MAKES THEM SUBJECT TO MORE FRICTION. IN ADDITION, THEIR LARGE BODIES, MADE LARGER BY PADS, ARE SLOWED DOWN BY AIR RESISTANCE. IF YOU WATCH CAREFULLY, THOUGH, YOU CAN SEE PLAYERS TAKING ADVANTAGE OF THEIR INERTIA. THEY WILL PUSH OFF AND GLIDE, IN A STRAIGHT LINE, ON THEIR WAY TO THE BENCH OR THE PENALTY BOX.

PHOTO: pointnshoot/creativecommons.org

motion, the puck tends to move at a constant speed and in a straight line. In fact, scientists have calculated that if a puck was launched in a very long rink at 160 kilometers per hour (100 miles per hour), it would only come to rest after traveling almost two kilometers (1.2 miles) in a little over two minutes. It is no wonder that it's difficult to notice a puck slowing down in the length of an ice rink.

Ice hockey is all about the speed and the direction of the puck—which is made possible by the icy surface of the rink. Players and announcers of hockey games speak of "fast ice," which has less friction, and "slow ice," which has more friction. Very cold ice is fast. As a hockey game progresses, "snow" is shaved from the ice by skate blades and hockey sticks. By the end of a period, the ice is full of gashes and grooves, and a dusting of snow covers the surface. The additional friction of the "snow" and the imperfections on the ice make the puck lose speed as it travels from stick to stick or from end to end. Because of this additional friction, many players handle the puck differently at the beginning of a period than at the end.

Careful attention is paid to restore the smoothness of the ice between periods. An ice

▶ IF YOU HAVE BEEN TO AN ICE HOCKEY GAME, YOU HAVE PROBABLY SEEN A ZAMBONI® ICE RESURFACING MACHINE IN ACTION.*

PHOTO: © Zamboni Company, David Klutho

*ZAMBONI and the configuration of the Zamboni® ice resurfacing machine are registered trademarks of Frank J. Zamboni & Co., Inc.

rink is about an inch thick, built from layer upon layer of water frozen at about -9 degrees Celsius (16°F), way below the freezing point of water. Once built, the ice is maintained by resurfacing machines. Before and after every game and between periods the ice is resurfaced to reduce friction. Ice resurfacing machines scrape off the "snow," remove dirt and debris, and then lay down a layer of hot water. The hot water melts the top layers of the ice and erases imperfections. After this process, which takes about twelve minutes from start to finish, the new surface is frozen and the ice is ready. Ice that has just been resurfaced is fast.

On the other hand, ice that is frozen and refrozen, like the ice of rinks in warmer climates, is slow. It seems the melting and freezing move more dirt and debris to the top layer. Ice that has melted and has a watery film on its surface is slow as well. Can you figure out why? ■

IN COLD TEMPERATURES, ICE CAN FORM ON ROADS AND WALKWAYS AND CAUSE SLIPPERY CONDITIONS FOR TRAVELERS.

PHOTO: Shea Hazarian/creativecommons.org

WHY IS ICE SLIPPERY?

Scientists continue to study what makes ice slippery. They all agree the top layer of ice is responsible. Conventional theories are that friction or the pressure of an object, like a hockey puck or skate blade, melts the top layer of the ice on contact. The water created by this melting serves as a lubricant, and objects glide on the water they generate.

More recent studies suggest that there are quasi-fluid, or water-like, layers on all ice, which cause it to be slippery. These slippery layers are not caused by pressure or by friction, but by the motion of the frozen water molecules. In quasi-fluid layers, the water molecules move up and down only, not in all directions, as is the case with water.

In very cold ice—ice at about -157 degrees Celsius (-250°F)—the slippery layer is very thin. It is only the depth of one molecule. As ice warms, there are more quasi-fluid layers. Hockey pucks and players have to cut through these layers as they travel across the rink. Could this explain why colder ice is faster?

DISCUSSION QUESTIONS

1. What conditions make for the best ice on a skating rink? Why do you think so?

2. What examples from your daily life demonstrate Newton's law of inertia?

EXPLORATION ACTIVITY: GETTING AROUND

A HOVERCRAFT IS A UNIQUE TRANSPORTATION DEVICE THAT TRAVELS OVER THE SURFACE OF THE WATER, UNLIKE A BOAT, WHICH TRAVELS THROUGH IT.

PHOTO: Universal Hovercraft, www.hovercraft.com

INTRODUCTION

The Exploration Activity is a research project that will give you the opportunity to apply what you learn in this unit. With a partner, you will select a single transportation device to study. You will do research to obtain as much information as you can about this transportation device during the next few weeks. You will focus on how this device demonstrates the concepts and principles of forces, energy, and motion that you have studied. You will share what you learn by making a poster or media presentation about the transportation device and by giving an oral presentation to the class.

OBJECTIVES FOR THIS LESSON

Understand the goals of the Exploration Activity.

Select a transportation device to research.

Develop a plan to research the selected device.

Work as team to research information.

Share knowledge with classmates.

▶ **MATERIALS FOR LESSON 8**

MATERIALS FOR BEGINNING THE EXPLORATION ACTIVITY

For you

1   copy of Inquiry Master 8.1: Calendar for the Exploration Activity

1   copy of Student Sheet 8.1: Guidelines for the Exploration Activity

1   copy of Student Sheet 8.2: Getting Started on the Exploration Activity

GETTING STARTED

1 Read "Civil Engineering: Danelle Bernard's Bridge to the Future" on pages 90-93.

2 Participate in a class discussion about what you have read.

3 Discuss with the class the importance of working together as a team.

PART 1

BEGINNING THE EXPLORATION ACTIVITY

PROCEDURE

1 Participate in a class discussion of the purpose and goals of the Exploration Activity.

2 To get started, you need to come up with some ideas for transportation devices that you can research. With the class, brainstorm the kind of devices you could investigate. Suggest any ideas that come to mind. Do not judge an idea right now. Do not worry if it is a good or bad idea. You will decide that later. You want to make a list of as many ideas as you can.

3 With the class, evaluate each device on the list. Some devices will be reasonable to research, others will not. Your teacher will delete from the list any devices that will not work.

▶ **WHAT KEY DESIGN FEATURES ALLOW THIS VEHICLE TO FLY?**

PHOTO: National Science Resources Center

4 With your partner, decide on the device you would like to investigate. If you would like to investigate a transportation device that is not on the list, check with your teacher to be sure that it is acceptable. Remember, it should be something that people use to travel from one place to another.

5 In your science notebook, record the transportation device you choose. 🔗

6 Your teacher will give you Student Sheet 8.1: Guidelines for the Exploration Activity. It has guidelines for completing your research. Follow along as your teacher reviews each part of the project. Make sure you understand what is expected for each part. Ask questions about anything that is not clear.

7 Now review the calendar on Inquiry Master 8.1. It tells you the dates by which each part of your project should be completed. Put the calendar in the front of your science notebook so you can refer to it later. Be sure to follow the calendar so that you complete everything on time and receive full credit for your work.

8 Discuss with your lab partner where you think you can find information about your device. Share your ideas with the rest of the class.

9 Complete Student Sheet 8.2: Getting Started on the Exploration Activity.

10 Read "How To Succeed With Your Project" on the next page.

READING SELECTION

BUILDING YOUR UNDERSTANDING

HOW TO SUCCEED WITH YOUR PROJECT

Research takes time and succeeds with steady, continuous effort. Your final grade for the Exploration Activity will be the sum of all the scores earned throughout the project, not a single grade given at the end. To earn the most credit, complete each part on time and follow directions.

You will complete the Exploration Activity over the next few weeks. You will do most of the work for it outside the classroom. Sometimes you will work on your own. At other times you will work with your partner at home, at school, or at the library. If you live near your partner, you will have opportunities to meet outside of school time to complete your research and design your presentation. Some class time will be given for working on the project. Make good use of it. It gives you a chance to ask any questions you may have.

A good plan of action will also help you complete the project. Your plan should include not only what you and your partner will do, but also when and where you will meet to complete your research and design your presentation. In addition to a plan, you will need a calendar with a schedule for completion of the project. Setting deadlines for different parts of the Exploration Activity makes it less likely that you will have to do a lot of work at the last minute.

Keep a log or journal of your work in your science notebook. The log should tell what you did to contribute to the group effort and when you did it. Record your thoughts and ideas as you work. You may want to have a parent sign your log. This shows you have been working steadily on the project.

Information comes in many forms. Learn to use all the different resources at your school or local library. Your teacher may arrange for you to go to your school library or computer lab to do some of your research. Librarians and computer resource teachers can help you find information and plan your presentation.

Think of different ways you can share what you learned. Putting the information on a poster is one way; using computers is another. Your teacher will give you information about how to plan your presentation. Choose the format that best fits the resources in your school.

Plan and practice your oral presentation. A well-organized presentation is the best way to get your information across. It will earn you the most credit. ■

PART 2

CONDUCTING YOUR RESEARCH

▶ **MATERIALS FOR CONDUCTING YOUR RESEARCH**

For you

1   copy of Student Sheet 8.3: Guidelines for the Poster Presentation

1   copy of Student Sheet 8.4: Guidelines for the Oral Presentation

PROCEDURE

1 Listen as your teacher gives you directions on using the media center or computer lab to research information about your device.

2 Use the resources of your media center and/or computer lab to find information about your device. You will use this information to make a poster or media presentation about your device and to plan an oral presentation of how the device works.

3 Read through Student Sheets 8.3 and 8.4. These sheets describe how your posters and oral presentations will be evaluated. Ask questions about anything that is not clear to you in the rubric.

4 When you have finished your research, design a poster about your device. Remember to refer to Student Sheet 8.3: Guidelines for the Poster Presentation to be sure your poster will have all the required elements on it.

PART 3

SHARING WHAT YOU'VE LEARNED
▶ **MATERIALS FOR SHARING WHAT YOU'VE LEARNED**

For you

Your copy of Student Sheet 8.4: Guidelines for the Oral Presentation

1   copy of Student Sheet 8.5: Exploration Activity Self-Assessment

PROCEDURE

1 Now you will share what you learned with your classmates and find out what they learned as well. You and your partner will display your poster or computer-generated presentation. Follow your teacher's directions for displaying your poster or computer-generated presentation. Keep in mind that you may want to use this display as part of your oral presentation.

2 Your team will also give an oral presentation to the class. Sharing knowledge is a key element of the scientific process. Your teacher will organize the order of the oral presentations for your class. Make sure you have everything you need for your presentation. Review Student Sheet 8.4, which describes the guidelines for the oral presentation, so you will know how your teacher will be assessing your presentation.

3 Listen to your classmates' presentations. As you listen, think about how what they did was like what you did. Also think about how what they did was different from what you did. What have you learned from their presentations?

4 After all the presentations, complete Student Sheet 8.5: Exploration Activity Self-Assessment.

Civil Engineering: Danelle Bernard's
Bridge TO THE Future

"When I was in high school," recalls Danelle Bernard, "I really liked math, science, and physics. I also was interested in architecture. For a while I thought about becoming an architect. But then I decided I wasn't 'artsy' enough."

Danelle's guidance counselor gave her a good idea. "Why not think about becoming a civil engineer?" he asked.

Danelle looked into civil engineering and soon decided that it was the career for her. She graduated from college with a Bachelor of Science degree in civil engineering. Her courses covered topics such as design of steel structures, design of concrete structures, soil mechanics, surveying, structural analysis, and construction cost estimating. After working for four years, Danelle took a test that qualifies those who pass it as licensed professional engineers. Many employers require licensing, which is like a degree.

Danelle eventually became a project engineer for the Bridge Design Division, which is part of the State Highway Administration in Maryland. Danelle and her co-workers have a big responsibility: to oversee the design of bridges throughout the state. The group is responsible for designs aimed at repairing old bridges (some of which are almost 100 years

▶ **DANELLE BERNARD**

PHOTO: Marvin D. Blimline, Maryland State Highway Administration

▶ BRIDGE MAINTENANCE IS THE
JOB OF CIVIL ENGINEERS.

PHOTO: Marvin D. Blimline, Maryland State
Highway Administration

2,500 bridges, and federal law requires that each bridge be inspected every two years.

The inspection teams examine each bridge carefully. If they see cracks or other signs of deterioration, the bridge is slated for repair work. The bridge inspection group has several teams of engineers who do the design for minor bridge repairs. However, if a bridge requires major repairs or has to be replaced because it is severely deteriorated, it is turned over to the Bridge Design Division, which includes Danelle's team.

The bridge design team carefully analyzes the forces acting on the bridge. Their objective is to determine how strong the structural members of the bridge have to be to carry the weight of the vehicles that will pass over it. They must make sure that the bridge meets national and state design codes, but at the same time they have to think about cost constraints. When you've got more than 2,500 bridges to think about, saving money is important!

Protecting the environment is also a concern. When a new bridge is being designed or an existing bridge is being repaired, Danelle and her team often meet with local citizens, elected officials, and members of environmental groups to discuss the effect that the bridge will have on the local community and the environment.

After Danelle and her colleagues have figured out all the details related to the bridge design, it's time to bring in the drafters, who transfer the designs into plans that will be used to build the bridge. In the past, drafting was painstaking work. A lot of time was spent drawing, erasing,

old) as well as building new ones. Some of the bridges are quite small. "They're out in the middle of a cornfield," she says with a smile. Many others, however, are large, concrete-and-steel structures located in urban areas. Thousands of vehicles pass over them every day. At any moment, Danelle and her group are working on about a dozen projects. Each takes about one to two years to complete.

TEAMWORK

One thing Danelle likes best about her job is that it involves teamwork. Several different groups of people, composed mainly of engineers, work on Maryland's bridges. One group, for example, is in charge of inspecting all of the state's bridges. The State Highway Administration is responsible for more than

▸ TODAY DRAFTERS USE COMPUTER TECHNOLOGY TO
REDESIGN BRIDGES.

PHOTO: Marvin D. Blimline, Maryland State Highway Administration

and drawing again! Today's drafters use CADD, short for "computer-aided drafting and design," to create the bridge plans. Changes can be made much more easily. In fact, using CADD, the drafters can even superimpose a drawing of a bridge onto a photograph to show how a new bridge will look in a certain area. That comes in handy when the bridge design team is working with local residents who are concerned that the bridge will destroy the appearance of their neighborhood.

ALWAYS SOMETHING DIFFERENT

Why is Danelle so enthusiastic about her work? There are many reasons. The first is simple—it's a job that is definitely needed. "Bridges help people get to work, get around, and do what they need to do!" she exclaims. Another big advantage is that she can see the results of her work. "Every time I drive on the highway," she says, "I can say to myself, 'I helped design that bridge.'"

The third reason is that every day brings new challenges. "Every bridge is different," Danelle says. "Every project goes differently. We don't just sit at a desk and 'crunch numbers.' We get involved with local citizens and people from diverse backgrounds. And when the bridge is under construction, we go out and meet with the construction crews. You're gaining new experience all the time. It's something that you could never learn from a textbook." ∎

DISCUSSION QUESTIONS

1. What training and skills does a person need to become a civil engineer? What other jobs utilize these skills?

2. What are some of the factors that might crop up during the planning process and delay the construction of a bridge?

THE FAN CAR

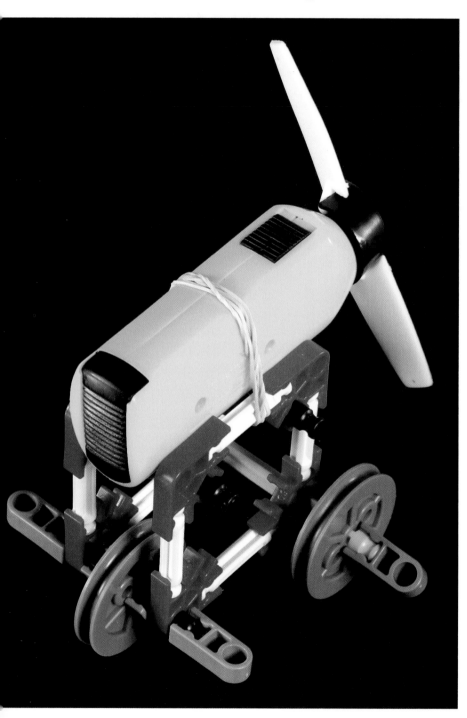

THE ASSEMBLED FAN CAR

PHOTO: © Carolina Biological Supply Company; used with permission

INTRODUCTION

In this lesson, you will continue your study of motion and forces. You will analyze the motion of a fan car you construct using K'NEX® parts and a battery-powered fan. You will observe the motion of the fan car with the fan turned off and with the fan turned on. In Lesson 10, you will study the motion of a vehicle powered by a mousetrap. Then, in Lessons 11 and 12, you will build a model roller coaster and study the motion of a car moving on the roller coaster track. As in previous lessons, you will make predictions, record observations, gather data, and draw conclusions based on evidence from your observations and data.

OBJECTIVES FOR THIS LESSON

Describe the force exerted by a battery-powered fan.

Describe the motion of a fan car.

Determine the effect of a constant unbalanced force on the speed of a fan car.

Calculate a fan car's average speed at different times as it moves along a path.

▶ **MATERIALS FOR LESSON 9**

For you

1	copy of Student Sheet 9.2: How Fast Is the Car Going?

For your group

1	battery-powered fan
2	AA batteries
1	rubber band
1	student timer
1	meterstick
1	2.0-m piece of adding-machine tape
1	20-cm piece of masking tape K'NEX® parts for fan car (see Appendix A: Directory of K'NEX® Parts):
8	gray connectors (C1)
8	red connectors (C4)
8	white rods (R2)
4	blue rods (R3)
1	yellow rod (R4)
3	small wheels (W1)
3	small tires (T1) (optional)

GETTING STARTED

1 Review how to find the average speed of a moving object.

2 If a car moves 200 meters in 40 seconds, what is its average speed?

3 Assemble the fan car as shown in Figures 9.1 and 9.2. (Your teacher will provide the batteries for the fan later in the inquiry.) Figure 9.1 is an exploded diagram, which shows you the parts needed to make the fan car. Figure 9.2 shows how the parts look in the completed car. Your teacher has a model fan car that you can also examine.

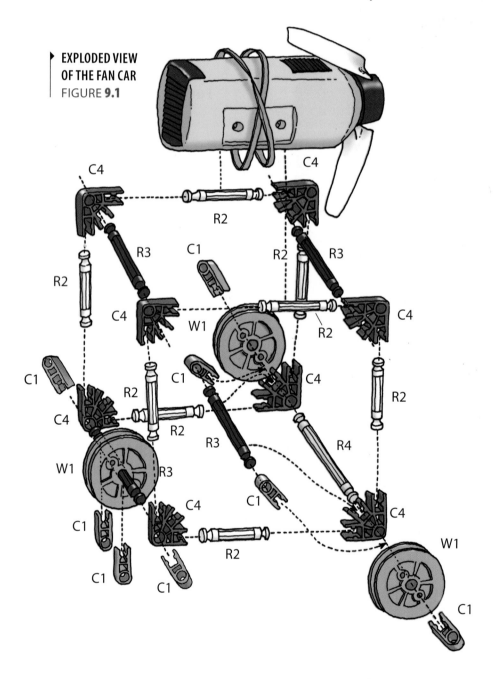

▸ **EXPLODED VIEW OF THE FAN CAR**
FIGURE **9.1**

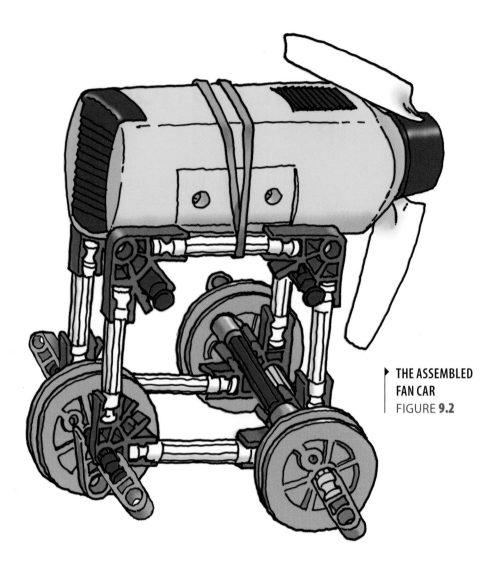

▶ THE ASSEMBLED
FAN CAR
FIGURE **9.2**

4 When you are finished, your car should look like the one in the photo at the beginning of this lesson. Check to make sure that it does.

5 What do you think the motion of the fan car would be if you put batteries in the fan and turned the fan on? Discuss your ideas with your lab partner.

INQUIRY 9.1

INVESTIGATING THE MOTION OF THE FAN CAR

PROCEDURE

1 Now that you have constructed your fan car, you will first make some predictions and observations about its motion with the fan off. In your science notebook, design a table to record your predictions and observations as you complete the activities below. Later, you will discuss your predictions and observations with the class. ☞

2 Record your predictions for the motion of the fan car if you push it and release it without the fan running.

3 Now push the fan car. Use a steady push. Record your observations of its motion after you release it.

4 Repeat Step 3 using forces of different strengths.

5 What differences in motion do you see when you change the push on the car? Cite evidence from your observations to support your answer.

6 When you pushed on the car, your hand exerted a force on it. List any other forces that were acting on the car when you pushed it.

7 Record the force(s) that acted on the car after you released it.

SAFETY TIP

Keep your fingers away from the moving fan blades.

8 You will now investigate the motion of the car with the fan on. Put two AA batteries in the fan base. Place the fan car on the table or on another flat surface, hold it in place, and turn on the fan without letting go of the car. Both you and your partner should do this. Answer the following questions:

A. What do you feel when you simply hold the car with the fan running?

B. In what direction does the fan move the air?

C. In what direction does the fan car want to move?

9 Before releasing the car, discuss with your partner how you think the fan car will move if you release it with the fan turned on.

10 Now release the car with the fan turned on. Observe and describe its motion.

11 Discuss with your partner how the motion of the fan car with the fan running compares with its motion after you released it with the fan turned off.

INQUIRY 9.2

MEASURING THE FAN CAR'S SPEED

PROCEDURE

1 With the class, review the behavior of the fan car turned off and turned on.

2 In this inquiry, you will measure the motion of the fan car. One way to do this is to measure the speed of the fan car as it moves across the table or floor. Review the information in the reading selection "Measuring Motion" on page 73. Find the equation for calculating average speed and write it in your science notebook. You will use the speed equation in this inquiry. ☞

3 Accurate timing is important for this inquiry. The times you measure will be very short. Before you begin collecting your data, practice your timing skills. Take turns operating the student timer and see who in your group is best at measuring short time intervals with it. Your teacher will provide an object for you to practice with. Drop the object and measure the time it takes the object to fall to the floor. The best timer is the person who can easily operate the student timer and get very close to the same timings for each drop. The best timer should operate the student timer for your group.

4 Place a long piece of adding-machine tape across the tabletop or floor. Use a piece of masking tape to mark a starting point at one end of the adding-machine tape.

5 Beginning at the starting point (0.0 m), mark distances in 0.4-m segments along the adding-machine tape, going all the way to 2.0 m (if possible), as shown in Table 1 on Student Sheet 9.2: How Fast Is the Car Going?

6 Label your tape distances along the tape as shown in Figure 9.3.

Adding-Machine Tape Measurements

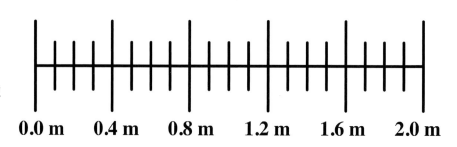

▶ HOW TO MARK THE TAPE FOR MEASURING AVERAGE SPEED OF THE FAN CAR

FIGURE 9.3

0.0 m   0.4 m   0.8 m   1.2 m   1.6 m   2.0 m

7 Use the student timer and determine the average time it takes the fan car to travel from the starting point to each distance you marked along the tape. Work as a team. It is important to start and stop the student timer at the right instants. Try to keep the car traveling in a straight line along the tape.

8 Use the average time to reach each distance on the tape to calculate the average speed of your fan car for each interval in Table 1 on Student Sheet 9.2. Calculate and record the speeds in Table 1 and answer the questions.

9 What patterns do you observe in your speed data? Answer this question in your science notebook. 📝

10 Follow your teacher's instructions to disassemble your car and return the parts to storage.

REFLECTING
ON WHAT
YOU'VE DONE

1 Answer the following questions in your science notebook. Be prepared to discuss your answers in class.

A. What are the forces on the fan car when the car is moving with the fan turned off and when it is moving with the fan turned on? What evidence do you have to support your answer?

B. Is the force of the fan constant or changing as the car moves along? Give reasons for your answer.

C. What is the effect of the force of the fan on the speed of the car? Cite evidence for your answer.

D. What can you conclude about the effect of a constant fan force on the motion of the car?

E. What energy changes take place as the car moves along with the fan running?

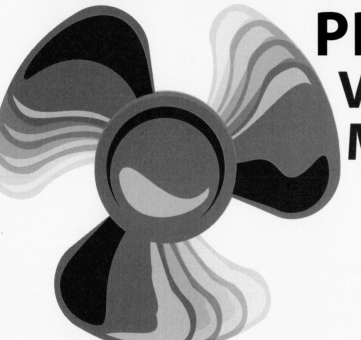

PROPELLERS: VEHICLES IN MOTION

In this lesson, you observed the spinning blades of the fan producing a force that moved your fan car. That fan acted like a propeller. Like all propellers, it had blades that rotated around a central hub. Fan cars aren't the only vehicles that use propellers. Vehicles for air, sea, and land all use propellers. ■

ON LAND

▶ IN 1931, TED JAMESON DESIGNED A MOTORCYCLE THAT UTILIZES THE POWER OF A PROPELLER. REFERRED TO AS THE WIND WAGON, THE ENGINE DRIVES AN AIRPLANE PROPELLER. THIS IS A UNIQUE FEATURE SINCE MOST MOTORCYCLE ENGINES OPERATE BY DRIVING THE VEHICLE'S WHEELS.

PHOTO: Courtesy of Lane Motor Museum, Nashville, TN

IN AIR

▶ HELICOPTERS USE THEIR PROPELLERS TO CREATE A FORCE THAT LIFTS THEM OFF THE GROUND. BY CHANGING THE ANGLE OF THE MAIN PROPELLER, THE PILOT CAN MAKE THE HELICOPTER MOVE FORWARD. THE SMALLER PROPELLER ON THE TAIL OF THE HELICOPTER CREATES A FORCE THAT KEEPS THE BODY OF THE HELICOPTER FROM SPINNING IN CIRCLES. ITS FORCE CAN ALSO BE USED TO TURN THE HELICOPTER.

PHOTO: NASA Ames Research Center

▶ MANY AIRPLANES HAVE PROPELLERS. LIKE THE PROPELLERS ON A HELICOPTER, THESE PROPELLERS ARE USED TO MOVE THE AIRPLANE FORWARD. HOWEVER, IN MOST PLANES, THE PROPELLERS DON'T DIRECTLY CAUSE THE PLANE TO LIFT OFF THE GROUND. THE ANGLE OF THE WINGS AND THE AIR FLOWING AROUND THEM LIFT THE PLANE INTO THE AIR. BUT A SPECIAL KIND OF PLANE, CALLED A TILT-ROTOR PLANE, HAS A PROPELLER THAT CAN BE USED LIKE A HELICOPTER'S PROPELLER TO LIFT THE PLANE OFF THE GROUND. WHEN THE TILT-ROTOR PLANE IS ALOFT, THE PROPELLERS CAN BE TURNED AND USED LIKE PROPELLERS ON A REGULAR AIRPLANE.

PHOTO: NASA Dryden Flight Research Center

ON WATER

▶ SOMETIMES UNDERWATER PROPELLERS CAN CAUSE PROBLEMS. IN SWAMPS AND PLACES LIKE THE EVERGLADES, WHERE THE WATER IS SHALLOW AND PLANTS GROW ALONG THE SURFACE, A PROPELLER CAN GET TANGLED IN THE PLANTS OR CAUGHT ON THE BOTTOM. AIRBOATS ARE DESIGNED TO SOLVE THESE PROBLEMS. AIRBOATS WORK MUCH THE SAME WAY AS FAN CARS. A FANLIKE PROPELLER ON THE BACK OF THE AIRBOAT PUSHES AGAINST THE AIR TO MOVE THE AIRBOAT ACROSS THE SURFACE OF THE WATER.

PHOTO: John and Karen Hollingsworth/U.S. Fish and Wildlife Service

BOATS, LIKE AIRCRAFT, USE PROPELLERS TO MOVE. MANY BOATS USE PROPELLERS BELOW THE SURFACE OF THE WATER. THESE PROPELLERS—UNLIKE THOSE ON AN AIRCRAFT OR A FAN CAR—PUSH AGAINST WATER INSTEAD OF AIR TO MOVE THE VEHICLE FORWARD. SOME BOAT PROPELLERS ARE VERY BIG. NOTICE THE SIZE OF THE PROPELLER COMPARED WITH THE SIZE OF THE MEN IN THIS PHOTOGRAPH. THIS PROPELLER IS ON THE *USS WINSTON S. CHURCHILL*.

PHOTO: IS1 Holly Hogan, U.S. Navy

DISCUSSION QUESTIONS

1. What are some of the pros and cons of using propellers for propulsion in air, sea, and land?

2. What else are propellers used for? Invent another application.

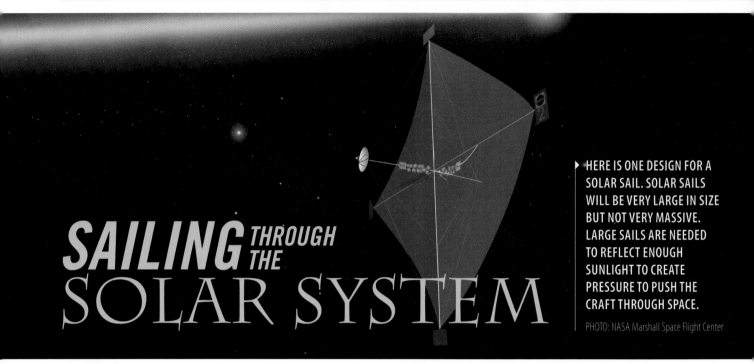

SAILING THROUGH THE SOLAR SYSTEM

▶ HERE IS ONE DESIGN FOR A SOLAR SAIL. SOLAR SAILS WILL BE VERY LARGE IN SIZE BUT NOT VERY MASSIVE. LARGE SAILS ARE NEEDED TO REFLECT ENOUGH SUNLIGHT TO CREATE PRESSURE TO PUSH THE CRAFT THROUGH SPACE.

PHOTO: NASA Marshall Space Flight Center

Scientists and engineers are always looking for ways to improve things. They try to achieve two goals: making things work better and making them cost less. The more complex a project, the bigger the challenge.

From this perspective, one of the biggest challenges is space travel. A great deal of energy is required to send spacecraft into space. Powerful rockets are needed to launch spacecraft and to provide the velocity they need to travel through the solar system. These rockets are expensive. The fuel is expensive, too.

How could space travel be made less expensive? One idea that scientists are exploring is solar power. Some scientists believe that a spacecraft could actually sail through the solar system. Solar-sailing spacecraft could travel large distances through the solar system using very little fuel.

SAILING BOATS ON EARTH

On the earth, sailboats glide across lakes, oceans, and bays. Sailboats need no fuel. They are powered by the wind. When the force of the wind is greater than the drag (friction) of a sailboat in the water, the wind pushes the boat forward. The navigator speeds the boat up or slows it down by controlling the angle the sails make with the wind. The navigator can also use the force of the wind and the rudder to change the direction the boat is going.

But in space, there is neither air nor wind. So how could a spacecraft sail? The answer is this: Spacecraft may be able to sail using the pressure of sunlight.

HOW SOLAR SAILS WOULD WORK

Sunlight is made of tiny energy packets called photons. If light traveling in straight lines out from the sun struck the sails of a spacecraft, the photons would bombard the sail like tiny ping-pong balls. When the photons struck the sail, they would push on it with a very small force. Because there is no air friction in space, the spacecraft would sail along with only the pressure of the photons of light and gravity acting on it.

So far, so good. But here's the drawback. A very large force is needed to accelerate something as big as a spacecraft. Photons come in vast quantities, but their individual power is small. One way to increase the force would be to make the sails of the spacecraft larger; in this way, more photons would hit it. But scientists have estimated that to catch enough photons to make a spacecraft move, the sails would have to measure a kilometer (0.6 mile) on each side. A sail that big would cover about 175 football fields!

MORE SOLUTIONS, PLEASE!

Another way to maximize the force of the photons on the sails would be to make sure the light reflects off the surface of the sails as strongly as possible. This could be accomplished by manufacturing sails from a material that reflects light like a mirror.

Another solution might be to make the sails as lightweight as possible. Scientists and engineers will be challenged to develop and design sails made of ultra-lightweight materials so the spacecraft will have as little mass as possible.

Scientists and engineers are creative. They have many ideas about how to make space sailing work. But they do agree on one thing: Designing huge sails that have very little mass and that are highly reflective is a real technological challenge.

There are no solar-sailing spacecraft yet, but scientists do hope to have them one day. These spacecraft could be placed in positions that would allow them to constantly monitor the earth and the sun. The findings they would transmit could help us better understand weather patterns and climate changes on the earth, as well as storms on the sun. Solar-sailing spacecraft might be able to visit planets, moons, comets, and asteroids and send back exciting new data. Some scientists even hope to use such vehicles someday to send a spacecraft to the stars.

Solar sailing is a dream of the future, but scientists and engineers are working to make it happen. New materials and technologies will be needed. New challenges will be met, and new discoveries made. ■

▶ **ON EARTH, SAILBOATS USE THE PRESSURE OF THE WIND.**

PHOTO: National Science Resources Center

DISCUSSION QUESTIONS

1. Compare and contrast solar sails and sailboat sails.

2. What challenges do scientists and engineers have to overcome to make solar sails work?

THE MOUSETRAP CAR

THE ASSEMBLED MOUSETRAP CAR

PHOTO: © Terry G. McCrea/Smithsonian Institution

INTRODUCTION

In Lesson 9, you built a fan car and measured its speed as it moved along the tape. In this lesson, you will build a mousetrap car and investigate its motion. You will design an experiment that will enable you to measure the speed of the car as it moves after the trap is released. You will also identify the forces acting on the car and describe how these forces affect the car's motion. You will then compare the motions of the fan car with those of the mousetrap car.

OBJECTIVES FOR THIS LESSON

Identify and describe the forces acting on the mousetrap car.

Observe and measure the speed of the mousetrap car as it moves.

Describe how forces affect the motion of the mousetrap car.

Describe the energy changes in the mousetrap car as it moves across the floor.

Compare the motion of the fan car with the motion of the mousetrap car.

▶ MATERIALS FOR LESSON 10

For your group

1	student timer
1	meterstick
1	0- to 10-N spring scale
1	piece of adding-machine tape
1	mousetrap
4	small washers
1	piece of string
1	piece of nylon line
1	piece of masking tape

K'NEX® parts for mousetrap car (see Appendix A: Directory of K'NEX® Parts):

6	gray connectors (C1)
2	tan connectors (C2)
14	red connectors (C4)
2	yellow connectors (C10)
6	white rods (R2)
5	yellow rods (R4)
4	red rods (R6)
2	small tires (T1)
2	large tires (T2)
2	small wheels (W1)
2	large wheels (W2)

GETTING STARTED

1 Assemble the mousetrap car as shown in Figure 10.1; the exploded diagram shows all the parts needed to assemble the car and how they connect. Figure 10.2 shows the car with the pieces properly connected. (The photo at the beginning of this lesson also shows the assembled mousetrap car.) It is important that you use a long piece of nylon line so that the axle will keep rotating after the trap has been released. If the nylon line is too short, it will unwind and then begin winding the opposite way.

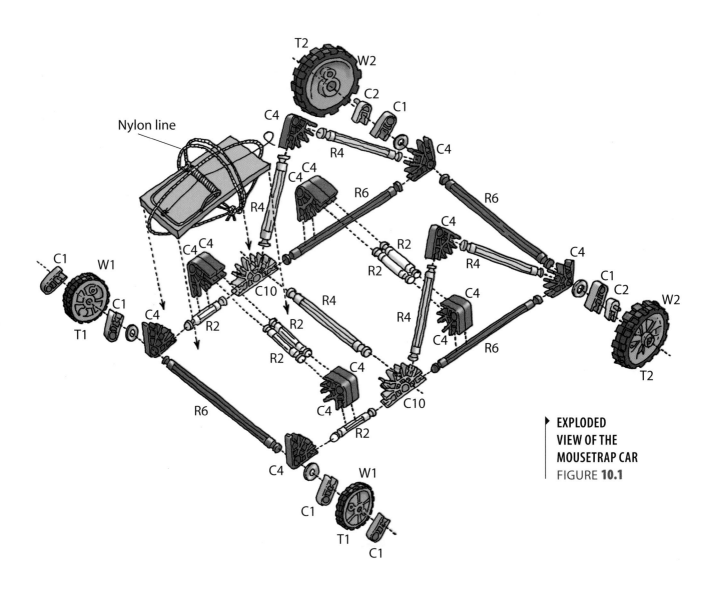

▶ **EXPLODED VIEW OF THE MOUSETRAP CAR**
FIGURE **10.1**

ASSEMBLED MOUSETRAP CAR. THE
MOUSETRAP IN THIS ILLUSTRATION
HAS BEEN SPRUNG. THE NYLON LINE
IS ATTACHED TO THE JAWS OF THE
TRAP AND TO THE REAR AXLE. BY
TURNING THE WHEELS, YOU CAN
WIND THE NYLON LINE AROUND THE
AXLE AND PULL THE JAWS OF THE
TRAP OPEN.
FIGURE **10.2**

SAFETY TIP

Do not put your fingers
in the clamping device of
the mousetrap.

2   Double-check your vehicle to be sure
that the trap is securely attached to the
car and that the jaws of the mousetrap
will open and close properly.

3   How can you put energy into the
mousetrap car? Discuss this question
with the class.

INQUIRY **10.1**

OBSERVING THE MOTION OF THE MOUSETRAP CAR

PROCEDURE

1 Complete the activities that follow. Record your observations and answers to the questions in your science notebook. Be prepared to share your observations with the class. 📝

2 Set the car on the table or floor and attach a spring scale to the mousetrap bar. Holding the car firmly, slowly pull the bar back with the spring scale (see Figure 10.3). Observe the force on the spring scale as you pull the bar. Record what happens to the force as you keep pulling the bar. Slowly release the bar so that it returns to its resting position, then remove the spring-scale hook from the bar.

3 Hold the car so that it does not touch the floor. Set the mousetrap spring by turning the wheels of the car so that the nylon line winds around the axle and pulls the mousetrap bar all the way back. Let go of the wheels while still holding the car off the floor. Do this several times. Discuss what happens with your group.

4 What do you think the motion of the car would be if you set the mousetrap and released it with the car on the floor? Record your prediction.

5 Reset the mousetrap. Place the car on the floor and release it. Describe what happens.

6 Write a paragraph describing the motions of the mousetrap car after the trap was set and the car was released. Discuss with your lab partner what forces you think are producing the motions.

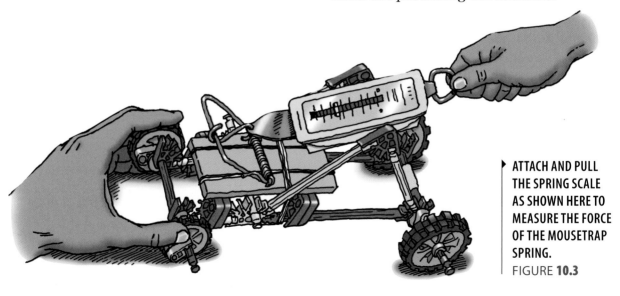

▶ ATTACH AND PULL THE SPRING SCALE AS SHOWN HERE TO MEASURE THE FORCE OF THE MOUSETRAP SPRING.

FIGURE **10.3**

MEASURING THE SPEED OF THE MOUSETRAP CAR

PROCEDURE

1 Share with the class what you wrote about the motion of your mousetrap car.

2 With the class, discuss the following questions:

A. Is the speed of the car constant as it moves across the floor?

B. How could you calculate the average speed of your car?

C. How could you design an experiment to measure the speed of the mousetrap car as it travels along the floor?

3 With your group, develop a plan to measure the motion. Design an experiment to determine the mousetrap car's speed at different positions along its path. Write your plan in your science notebook. Design a data table on which to record your measurements and any calculations. ☞

4 Carry out your plan.

5 When you are done, summarize the conclusions you can draw from your data.

6 Follow your teacher's instructions to disassemble your car and return the parts to storage.

REFLECTING
ON WHAT
YOU'VE DONE

1 Answer the following questions in your science notebook. Be prepared to share your answers with the class.

A. Summarize what you found out about the motion of the car. Support your conclusions with the data you collected.

B. Explain the changes in the mousetrap car's speed in terms of the forces acting on the car.

C. When you set the mousetrap, do you do work on it? If so, how?

D. Write a paragraph describing the energy changes that took place when you set the trap and released the car.

E. How does the motion of the mousetrap car compare with the motion of the fan car? Identify similarities and differences. What forces acted each time?

Medieval Warfare
in Modern Times

young man wearing leather armor and carrying a wooden shield runs from the fire of a catapult that is throwing missiles at him. Temporarily out of the catapult's range, he pauses to rest, checks his digital watch, and looks around at the Arizona landscape. Digital watch? Arizona? This man isn't a medieval warrior but a member of a war reenactment group in the 21st century. The catapult is built, as much as possible, like a medieval catapult. However, rather than throwing heavy stones, reenactment catapults throw groups of tennis balls that are taped together. (The word "catapult" comes from two Greek words: *kata* means "down" and *pallein* means "to hurl.")

Catapults were war machines during medieval times. They were used to attack castles and fortresses. The catapults hurled large stones and other things at the castle walls or even over the walls of fortresses. Persistent battering could eventually win the battle. Medieval warriors used at least three different kinds of catapults, and the people who participate in medieval battle reenactments today build and use all three of them.

One kind of catapult is the mangonel. The Romans designed it in the third century. It was the most popular kind of catapult of the medieval period. It is also the most popular catapult built for reenactments today. How does a mangonel work?

THE MANGONEL WAS A POPULAR TYPE OF CATAPULT DURING THE MEDIEVAL PERIOD. THE THROWING ARM WAS USED TO LAUNCH ROCKS THROUGH THE AIR.

Setting a mangonel is very much like setting a giant mousetrap. A mangonel has a single arm with a cuplike extension at the end. Two ropes attached to this arm can be wound around a pole using a lever. As the ropes are wound around the pole, the throwing arm is pulled down so that it can be loaded. The more tightly the rope is wound, the greater the force pulling on the throwing arm—the same as when you pull back the bar on a mousetrap. When the throwing arm is released, it snaps forward into a crossbar, which suddenly stops the throwing arm and sends a rock flying out of the cup and through the air. Unfortunately, this is an inefficient kind of catapult, since much of the available energy is lost into the framework of the catapult when the throwing arm hits the crossbar. Only a small portion of the energy put into the catapult when the ropes are stretched is converted to energy in moving the stone.

A second type of catapult from medieval times is called a traction trebuchet. With a long pole mounted on a tall frame, this catapult uses the principles of a lever. The pole is positioned so that the fulcrum is close to one end. A sling that holds a rock is attached to the end of the pole farthest from the fulcrum. Ropes are attached to the other end of the pole, which is at the end closest to the fulcrum. When a crew of warriors pulls down on these ropes at the end closest to the fulcrum, the long end of the pole rises quickly into the air and sends the rock hurtling toward the target.

In recent years, a crew of five people using a reconstructed traction trebuchet was able to throw a 900-gram (2-pound) lead ball 170 meters (558 feet). Medieval traction trebuchets were known to have crews of 30 men or more.

A much more powerful catapult used in medieval times was the counterweight trebuchet. Like the traction trebuchet, the counterweight trebuchet uses the principles of a lever. However, gravity, rather than a crew of warriors, provides the downward force that sends the rock into the air. A heavy weight is attached to the short end of the pole. The longer end has to be pulled down by the crew and loaded before it can be used. When the long end of the pole is released, gravity pulls the heavy weight on the short end down. The long end is raised into the air, and the stone is sent flying. This design

▶ TODAY PEOPLE PUT CATAPULTS TO NOVEL USES, SUCH AS LAUNCHING PUMPKINS.

PHOTO: Joe Shlabotnik/creativecommons.org

worked well; 44 such catapults spread havoc around Europe during medieval times. A modern counterweight trebuchet with a 5400-kilogram (11,905-pound) weight has been used to throw a 635-kilogram (1,400-pound) car 79 meters (259 feet) and 45 kilograms (99 pounds) of iron 215 meters (705 feet).

While catapults like these and others from medieval times are no longer used in war, they are still of great interest to a number of people. War-reenactment groups, historians, and others build them. Over the years, all sorts of items have been launched with catapults—from stones and spears to people, pianos, and pumpkins. ■

▶ THE TREBUCHET WAS A KIND OF CATAPULT. THIS ONE USED COUNTER-WEIGHTS TO FIRE ITS LOADS.

PHOTO : Artist conception by George L. Warfel, American Ordnance Association

DISCUSSION QUESTIONS

1. What forces may be at work in the use of a catapult?

2. Have you ever encountered a catapult in your life? What was it used for?

Rocket Science 101

You don't need to be a rocket scientist to understand how a rocket works. All you need to know is Newton's Third Law of Motion, which states that for every action, there is an equal and opposite reaction.

Newton's Third Law of Motion works for all forces, including the ones you have been exploring in the lab. For example, when your fan car was running, the fan pushed on the air and the air pushed back on the fan. It was the air pushing on the fan that pushed the car. The air went one way and the car went the other.

Something similar happens with rockets. Rockets go forward by expelling hot gas backward. Rocket engines generate the force to push the gas out at a very high speed by burning a lot of fuel quickly. Rapidly burning fuel creates the huge force that sends the hot gases out of the rocket. According to Newton's Third Law of Motion, as the rocket pushes the gas out the back, the gas pushes the rocket forward. Newton, in fact, knew that if something was launched with enough force, it could gain enough speed to orbit Earth or even to escape Earth, but he did not have rockets powerful enough to do it.

To launch a rocket, a rocket engine must expel its gases with enough force (thrust) to exceed the force of gravity. People had been using this principle for centuries. For example, the Chinese were using rockets for military purposes in the 13th century. The Chinese also invented fireworks, which are another example of rockets. These early rockets were powered by solid fuel, which was similar in composition to gunpowder.

In the early 1800s, a British officer, William Congreve, improved existing rockets for military use. Their glow inspired the words "rockets' red glare" in "The Star-Spangled Banner," the national anthem of the United States. But none of these rockets was powerful enough to launch anything into space. Like the Chinese rockets, the rockets launched with this gunpowder-type fuel all fell back to the ground.

Launching satellites into orbit and sending them to other planets requires very powerful rockets. The force of the rockets must push on the satellites and give them enough speed so they will go around Earth continuously and not fall back to the ground. To send a spacecraft to the moon or to other planets in our solar system, the force of the rocket has to be strong enough, and last long enough, to give the spacecraft the speed to escape Earth and not be bound by its gravity. The speed needed to leave Earth and not be bound by its gravity is called escape velocity. Clearly, if space travel were to become a reality, scientists would need to find a way to make more powerful rockets.

It wasn't until 1926 that American physicist Robert Hutchings Goddard developed and launched the world's first liquid-fueled rocket. This breakthrough eventually led to the development of rockets powerful enough to launch satellites and other spacecraft into orbit around Earth. In 1957, Russia successfully launched *Sputnik*, the first artificial satellite. In

1958, the United States successfully launched a satellite into space. It was named *Explorer I.* These satellites orbited Earth and were a wonder of science and technology. They marked the beginning of the Space Age.

In the 1960s, scientists and engineers in the United States developed and built *Saturn V,* the biggest and most powerful rocket ever built. Dr. Wernher von Braun, a German physicist, headed the team. He was brought to the United States to design rockets after working with the Nazis to build V-2 rockets during World War II, which were used to carry warheads. Those rockets were a key step towards the later development of rockets that could go to outer space.

DR. ROBERT GODDARD STANDS NEXT TO A LIQUID OXYGEN-GASOLINE ROCKET HE DESIGNED. THE ROCKET WAS FIRED FROM THE FRAME PICTURED ON MARCH 15, 1926 IN AUBURN, MASSACHUSETTS.

PHOTO: NASA Marshall Space Flight Center

The *Saturn V* stood almost 20 meters (65.6 feet) higher than the Statue of Liberty; on the launch pad, it weighed 13 times more than the statue. The engines on the rocket had the horsepower of 4,300 automobiles. *Saturn V* was used to send a three-man crew to the moon in 1969. It needed so much fuel that it was built in stages, or sections. Each stage was released when the fuel in it was used up. This process made the rocket lighter after each release so that it was easier to speed up the rocket and send its payload of astronauts on their way.

Rockets like the *Saturn V* are no longer being built. They are too expensive, and future manned missions would require rockets even more powerful than the *Saturn V*. But scientists and engineers continue to improve rocket technology. They are exploring new designs that will eventually enable humans to travel back to the moon and beyond. Someday, you will even be able to take a space vacation. (Book early. Reservations required.) ■

▶ **POWERFUL ROCKETS ARE NEEDED TO LAUNCH A SPACE SHUTTLE INTO ORBIT.**

PHOTO: NASA Kennedy Space Center

FROM 1968 TO 1972, THE APOLLO MISSION SENT CREWS OF ASTRONAUTS TO EXPLORE THE MOON. ASTRONAUTS TOOK THIS PICTURE, WHICH SHOWS HOW THE EARTH LOOKS AS SEEN FROM THE MOON.

PHOTO: NASA Headquarters – Greatest Images of NASA

DISCUSSION QUESTIONS

1. How are the forces in a rocket similar to the forces in a mousetrap car? How are they different?

2. What innovations were necessary to make a huge rocket like the *Saturn V* work?

THE ROLLER COASTER

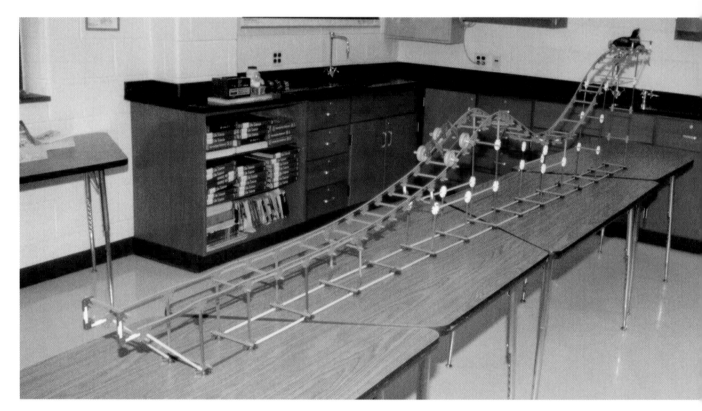

▶ **THE ASSEMBLED ROLLER COASTER**

PHOTO: © Terry G. McCrea/Smithsonian Institution

INTRODUCTION

In Lessons 9 and 10 you investigated the motion of a fan car and a mousetrap car. You looked at how the speed and kinetic energy of the cars changed as they moved across the floor. In this lesson, you will build a roller coaster and a roller coaster car. You will test the roller coaster to make sure it works properly. In the next lesson, you will learn how to put energy into the roller coaster car. You will then investigate how the speed and energies associated with the car change as it moves on the track.

OBJECTIVE FOR THIS LESSON

▶ Build a roller coaster.

▶ **MATERIALS FOR LESSON 11**

For the class

K'NEX® parts for the roller coaster
and roller coaster car (see Appendix A:
Directory of K'NEX® Parts):

4	gray connectors (C1)
12	red connectors (C4)
12	green connectors (C5)
30	purple connectors (C6)
30	blue connectors (C7)
100	gold connectors (C8)
22	white connectors (C9)
8	yellow connectors (C10)
50	green rods (R1)
22	white rods (R2)
46	blue rods (R3)
20	yellow rods (R4)
70	gold rods (R5)
42	red rods (R6)
40	gray rods (R7)
4	large wheels (W2)
2	4.25-m strips of K'NEX® track

GETTING STARTED

1 Look at the photo on page 120, which shows an assembled roller coaster. You and your classmates will build a roller coaster just like this one. To do it, you need to divide the work among the members of your class and work cooperatively. Discuss a plan for developing student groups to work on different parts of the roller coaster. Each group should identify the part of the roller coaster it will assemble.

2 Check to see that your group has all the necessary K'NEX® pieces for its section of the roller coaster. With your group, identify the illustrations in the Student Guide that show how to assemble your section of the roller coaster.

▶ **WHAT DOES THE ASSEMBLED K'NEX® MODEL HAVE IN COMMON WITH THE CONSTRUCTION OF AN ACTUAL ROLLER COASTER? WHAT IS DIFFERENT?**

PHOTO: Grantuking/creativecommons.org

BUILDING A ROLLER COASTER

PROCEDURE

1. Figures 11.1 through 11.6 show how to assemble and connect the sides of the roller coaster. Each student group will assemble and connect the two sides needed for each section. Figure 11.7 shows how to connect the two sides of the roller coaster using the gold rods (R5). Figures 11.8 through 11.10 show how to assemble and connect the completed sections. Use the illustrations that match the sections your group is assembling. The orange track shown in the illustrations is attached after all sections have been connected. Figures 11.11 and 11.12 show what the finished roller coaster looks like. Figure 11.13 shows how to assemble the roller coaster car and what the completed car will look like.

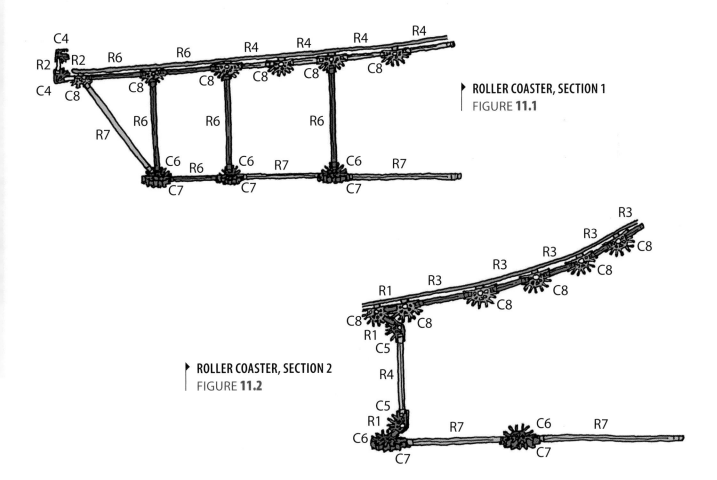

▶ ROLLER COASTER, SECTION 1
FIGURE **11.1**

▶ ROLLER COASTER, SECTION 2
FIGURE **11.2**

Inquiry 11.1 continued

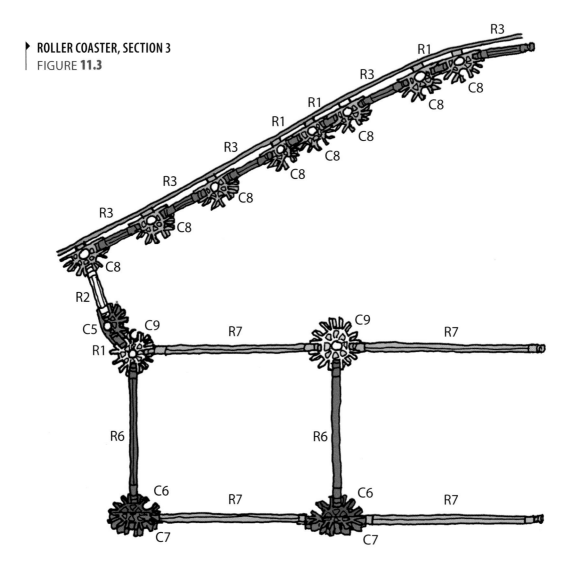

▶ ROLLER COASTER, SECTION 3
FIGURE **11.3**

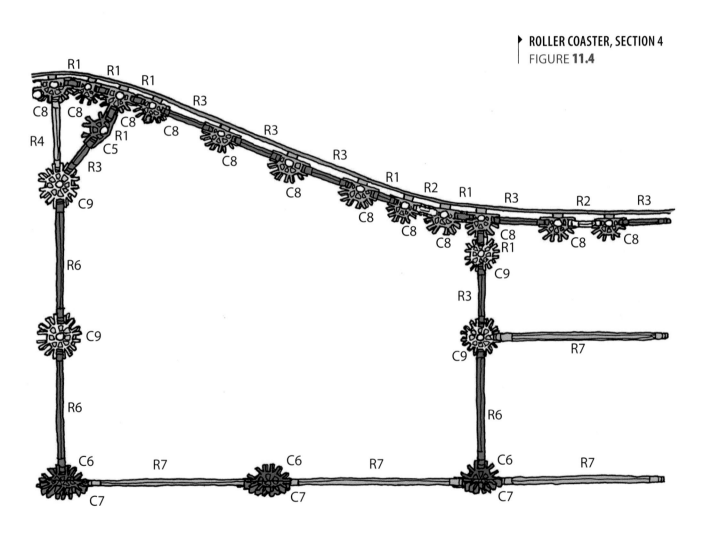

Inquiry 11.1 continued

▶ ROLLER COASTER, SECTION 5
FIGURE **11.5**

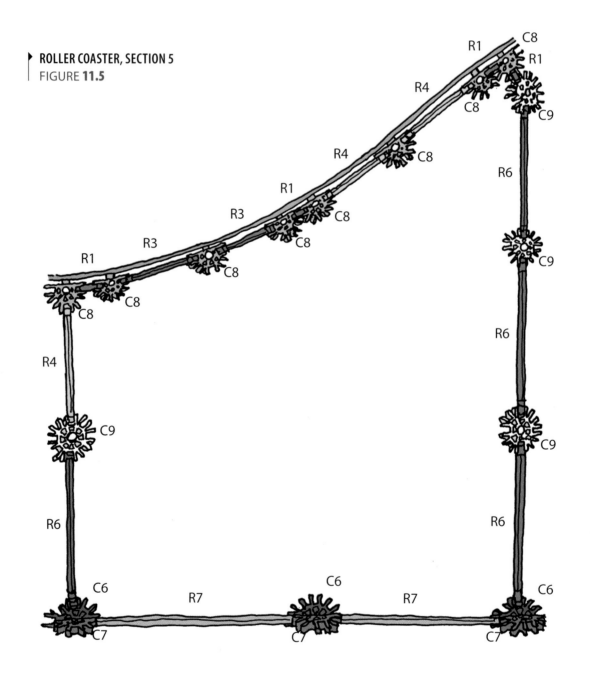

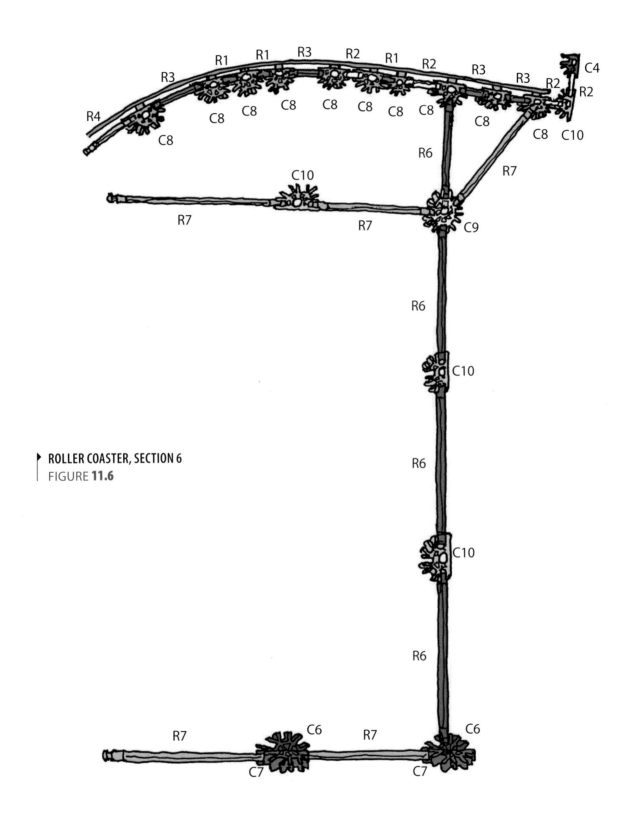

ROLLER COASTER, SECTION 6
FIGURE **11.6**

Inquiry 11.1 continued

▶ **HOW TO CONNECT GOLD RODS TO THE ROLLER COASTER**
FIGURE **11.7**

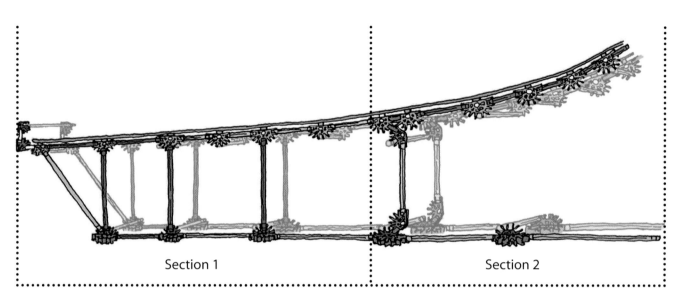

Section 1   Section 2

▶ ROLLER COASTER A (SECTIONS 1 AND 2 CONNECTED)
  FIGURE **11.8**

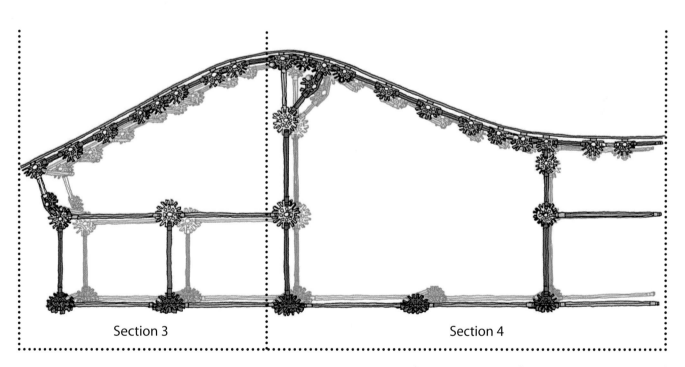

Section 3   Section 4

▶ ROLLER COASTER B (SECTIONS 3 AND 4 CONNECTED)
  FIGURE **11.9**

Inquiry 11.1 continued

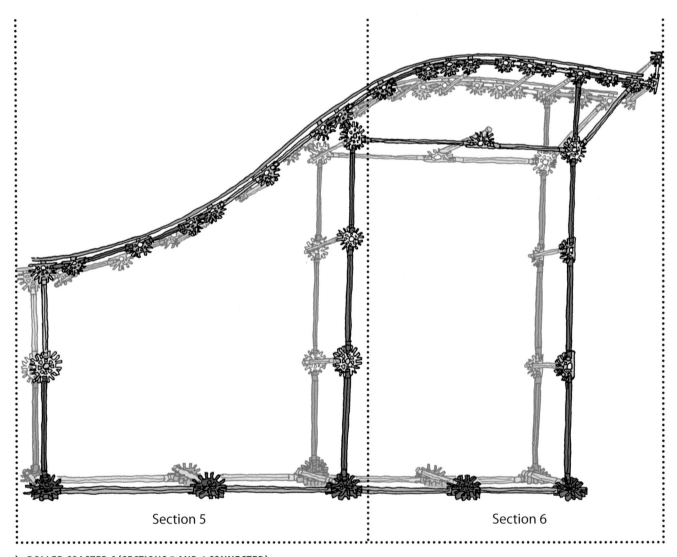

Section 5

Section 6

▶ ROLLER COASTER C (SECTIONS 5 AND 6 CONNECTED)
FIGURE **11.10**

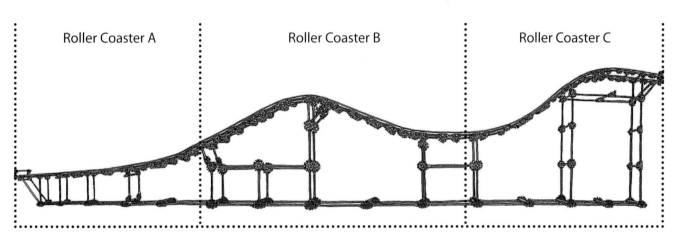

Roller Coaster A	Roller Coaster B	Roller Coaster C

▶ **FINISHED ROLLER COASTER, SHOWING A, B, AND C CONNECTED**
FIGURE **11.11**

Finished Roller Coaster

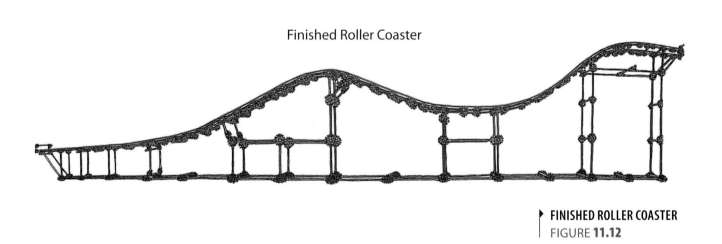

▶ **FINISHED ROLLER COASTER**
FIGURE **11.12**

Inquiry 11.1 continued

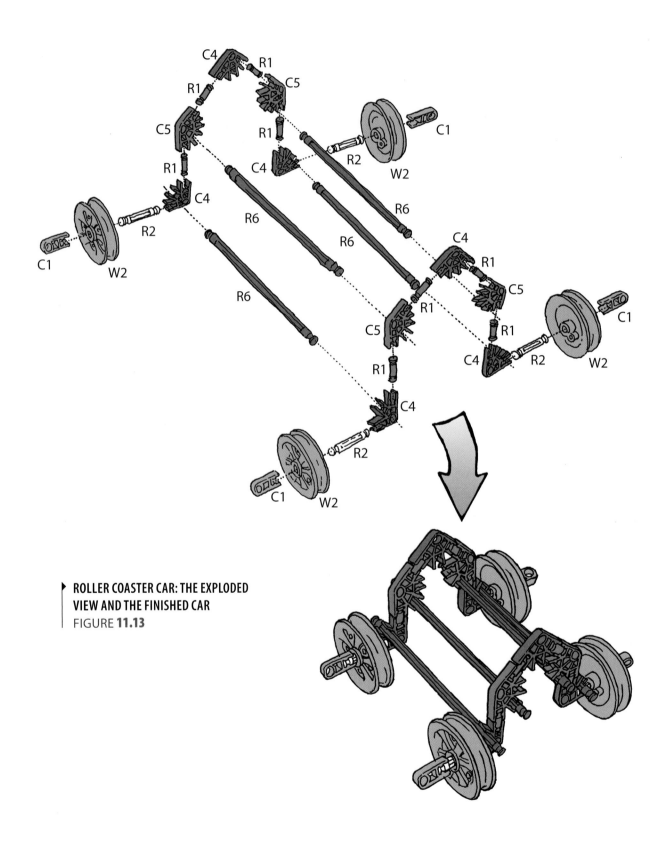

▶ **ROLLER COASTER CAR: THE EXPLODED VIEW AND THE FINISHED CAR**
FIGURE **11.13**

2 After each group has connected its section of roller coaster, the group responsible for the track should attach the track to each side of the roller coaster. The track must be stretched smoothly and tightly along the coaster frame. Appendix A has tips for putting the track on the roller coaster.

3 The group that built the roller coaster car should then place the car on the roller coaster at the high end of the track and test the car to make sure that it coasts smoothly and remains on the track for the entire length of the roller coaster. If the car does not move smoothly, check to see that all the parts are connected correctly and that the track is attached smoothly.

REFLECTING
ON WHAT
YOU'VE DONE

1 In your science notebook, describe your contribution to the class task of building the roller coaster.

2 Place the roller coaster car at the low end of the roller coaster track. Describe the motion of the car.

3 Discuss with the class ways to put energy in the car so that it will move along the track.

4 Discuss with the class how roller coasters in amusement parks get the energy they need to move along the track.

MOTION ON A ROLLER COASTER

▶ **THE LAWS OF PHYSICS IN ACTION!**

PHOTO: Cedar Point Amusement Park/Resort, Sandusky, OH

INTRODUCTION

When you enjoy the thrill of a ride on a roller coaster, you are experiencing the laws of physics and energy and motion in action. In this lesson, you will use the roller coaster and roller coaster car your class built in Lesson 11. You will investigate energy transformations as the roller coaster car moves along the roller coaster track. You will also compare the motion of the roller coaster car with the motions of the fan car and the mousetrap car that you built in previous lessons.

OBJECTIVES FOR THIS LESSON

Observe and describe the motion of the roller coaster car as it moves along the track.

Predict the motion of the roller coaster car when it is released at different points along the track.

Measure the speed of the roller coaster car at several points on the track.

Describe the energy changes in the roller coaster car as it moves along the track.

For the class

1	roller coaster
1	roller coaster car
1	0- to 10-N spring scale
1	meterstick
1	student timer
1	piece of masking tape

GETTING STARTED

1 If you have not read "Potential and Kinetic Energy," read it at this time.

2 In your science notebook, write your answers to the following questions. Your answers should be based on what you read in "Potential and Kinetic Energy." 🖎

A. What does it mean for something to have potential energy?

B. If a book weighs 15.0 N, what is its gain in gravitational potential energy when it is lifted onto a shelf 2.0 m above the floor?

C. How can you tell whether something has kinetic energy?

D. Do you ever have kinetic energy? How do you get it? How do you lose it?

3 Have your science notebook ready so you can write your predictions, observations, and answers in it as you perform the inquiry.

▶ **WHICH BOOK HAS THE MOST GRAVITATIONAL POTENTIAL ENERGY?**

PHOTO: Monterey Public Library/ creativecommons.org

READING SELECTION

BUILDING YOUR UNDERSTANDING

POTENTIAL AND KINETIC ENERGY

You have investigated energy and changes in energy throughout this unit. You have seen how energy can be stored in batteries and springs.

STORING ENERGY

When work is done on something, its energy changes. For instance, when you set the spring in the mousetrap car, you did work on the spring because you exerted a force on the spring when you wound it around the axle of the car. The spring gained energy and stored it for later use. Scientists call this stored energy potential energy.

When you let go of the car, the spring released the potential energy. The force of the spring did work on the mousetrap car and increased its speed. The stored energy in the spring became kinetic energy, or energy of motion. Some of the energy in the spring also became heat energy because of friction. Eventually, friction stopped the car, and all the kinetic energy of the car was transformed into heat energy.

The batteries you used in the fan car also stored energy. The batteries stored energy as chemical potential energy that became electric potential energy. When the batteries were connected to the fan, the electric potential energy was transformed into kinetic energy.

Energy can be stored in other ways, too. You can store energy in an object by lifting it. Whenever you pick up an object you do work on it because your muscles exert a force on an object through a distance. Your muscles changed the chemical energy in your body into energy in the lifted object. But what kind of energy is associated with the lifted object?

GRAVITATIONAL POTENTIAL ENERGY

The energy an object gains when it is lifted is called gravitational potential energy. It is called gravitational potential energy because you must work against gravity to lift an object to a higher position above the ground (earth). Any object above the ground has gravitational potential energy that can be released.

The amount of gravitational potential energy an object has depends on how much it weighs and how high above the ground it is. For example, it hurts more if a heavy book falls from your desk and hits your foot than if a light book falls and hits you from that same height. If you drop the same book from different heights, however, you will find that the higher the book's starting position, the more your foot hurts when the book hits it.

To calculate the gravitational potential energy of an object, you multiply its weight by its height above the ground, as shown in the following equation:

$$\textbf{GRAVITATIONAL POTENTIAL ENERGY} \quad = \quad \textbf{WEIGHT} \times \textbf{HEIGHT}$$

Weight is measured in newtons, and height is measured in meters. Therefore, the unit of measure for potential energy is newton-meters. Newton-meters are also called joules. Thus, the units for energy are the same as the units for work. Remember: energy is the ability to do work.

Kinetic energy is the energy of motion. An object gains and loses kinetic energy as it speeds up and slows down.

In this lesson, you will study how gravitational potential energy and kinetic energy change as a car travels along a roller coaster. You will learn how to put gravitational potential energy in the car, and you will observe how energy converts from one form to another as the car moves along the track. ∎

INQUIRY **12.1**

OBSERVING THE MOTION OF A ROLLER COASTER CAR

PROCEDURE

1 Discuss the following questions with your group. Then share your answers with the class.

A. What kind of energy would the roller coaster car gain if you lifted it to the top of the roller coaster?

B. How does it get this energy?

2 Before you place the roller coaster car on the track, make the following predictions about what the speed of the car would be if it were released from the top of the track:

A. At what point along the track will the car have the fastest speed? The slowest speed? Why do you think so?

B. Where will the car have its greatest kinetic energy?

C. How do you think the car gained its kinetic energy?

D. Is it possible for the car to have both kinds of energy anywhere on the track?

3 Place the car at the highest point on the roller coaster track. Let it roll down. Compare your observations of its motion with your predictions. Discuss this with your group.

4 Based on your observations and what you have learned about energy, discuss with the class why the car moves the way it does along the track.

5 Predict what the motion would be if you placed the car at the top of the lower hill and allowed it to move back toward the higher hill.

6 Test your prediction by observing the car's motion. Answer this question in your science notebook: Based on what you have learned about energy, how do you account for the car's motion when the car is released from the lower hill? Then discuss your ideas with the class. 🖉

MEASURING THE SPEED OF A ROLLER COASTER CAR

PROCEDURE

1 You will measure the roller coaster car's speed at different points along the track. Before you make speed measurements, consider these questions:

A. What force "pulls" on the car as it moves along the track?

B. Do any other forces act on the car as it moves along the track? If so, what are they? How do they affect the motion?

2 Use a spring scale to measure the weight of the roller coaster car.

3 How much work must you do to lift the car from the tabletop to the top of the highest hill on the roller coaster? Write out your calculation in your science notebook. 🔗

4 Working with your group, develop a plan to measure the speed of the car from at least three different points along the roller coaster track. Determine how the speed of the car changes as it moves along the track.

5 Carry out your plan. Be sure to discuss the following questions:

A. How did your group select points along the track? What methods did your classmates use?

B. How did your speed values compare with those of the other groups?

REFLECTING
ON WHAT
YOU'VE DONE

1 Write responses to the following questions in your science notebook and then discuss your answers with the class.

A. What have you learned about the motion of the roller coaster car as it moves along the track?

B. What energy changes took place as the car moved along the track?

C. What changes could you make in the roller coaster to make the roller coaster car go faster?

D. How does the motion of the roller coaster car compare with the motion of the fan car and the mousetrap car? How are these motions alike? How are they different?

Twists, Turns, and Loops

▶ EACH ROLLER COASTER USES DIFFERENT COMBINATIONS OF TWISTS, TURNS, DIPS, AND DIVES TO CONVERT GRAVITATIONAL POTENTIAL ENERGY INTO KINETIC ENERGY.

PHOTO: Cedar Point Amusement Park/Resort, Sandusky, OH

Kuh-chink, kuh-chink, kuh-chink. The train lurches slowly up the first hill of the roller coaster. You give the lap bar one last tug. Then you round the first corner, pick up speed, and … Aaaaaaggghh!

You may not be thinking about the fundamentals of physics while you're riding a roller coaster, but those fundamentals, especially the laws dealing with energy and motion, are what keep you in your seat. They ensure that your ride is safe—as well as fun.

A roller coaster is a fairly simple machine. A chain that is attached to a motor pulls a train of cars filled with people to the top of a steep hill. When the cars are released, the thrills begin!

ENERGY AND SUDDEN TURNS PROVIDE PLENTY OF THRILLS AS YOU RIDE A ROLLER COASTER.

PHOTO: Cedar Point Amusement Park/Resort, Sandusky, OH

At the top of the hill, a roller coaster train and its riders have a large amount of stored, or potential, energy. The work the motor did to drag its load against the force of gravity to the top of the hill stored energy in the train and riders as it pulled them to the top. The higher the hill and the more work the motor did, the more gravitational potential energy it stored in the train and riders.

When the train and riders crest the hill and begin to rush downward, their gravitational potential energy begins turning into kinetic energy. The shape of the track and the height of the hills control the train's changes in speed and direction.

As the train and riders descend farther and farther, more and more gravitational potential energy turns to kinetic energy. That means the train and riders go faster and faster. As they rise, the train and riders gain back some of their gravitational potential energy and lose kinetic energy and slow down. Tight curves, which provide sudden direction changes, add to the thrills.

The first roller coasters were made from wood. They were not as efficient in converting potential energy into kinetic energy as today's steel roller coasters are. Roller coasters made of steel do a better job because there is less energy lost to heat and more energy to be spent on thrills.

Other improvements have been made in roller coaster design. Most of today's steel coasters include at least one bit of looped track that momentarily turns your world upside down. The looped track exerts a force on the train cars and on you and the other riders that sends you around the loop. Many coasters have multiple loops and corkscrews to maximize the fun.

Finally, your ride is over. You've been up and down and all around—a very energetic experience indeed! ∎

▶ **GOING AROUND A LOOP ON A ROLLER COASTER. THE TIGHTER THE LOOP, THE HARDER YOUR SEAT PUSHES AGAINST YOU. WHY?**

PHOTO: Cedar Point Amusement Park/Resort, Sandusky, OH

DISCUSSION QUESTIONS

1. Will a roller coaster with a higher starting point be a faster ride? Why or why not?

2. What are some other examples of humans using motors to drag loads against the force of gravity?

ISAAC NEWTON GOES SKIING

We don't know whether Sir Isaac Newton ever tried skiing. It's entirely possible, because skis were invented before his birth in 1642. Skis have been in use for more than 2,000 years—long before Newton ever thought about gravity or came up with his laws of motion.

Even though Newton may never have made the connection between skiing and his three laws of motion, you can. In fact, knowing Newton's Laws of Motion is useful if you go skiing. Knowing about gravity helps too.

THE FIRST LAW

Take Newton's First Law of Motion: An object at rest will remain at rest, and an object in motion will remain in motion with the same speed and direction, unless acted upon by an outside force. That means it takes a force to start you moving and another force to make you stop. You also need to apply a force if you want to change the

direction in which you're moving. If no forces act on you, you just keep moving along at the same speed and in the same direction.

How does this apply to skiing? If you're skiing across flat terrain (cross-country skiing), you have to exert a force to get yourself moving. You do that by pushing with your poles. You keep moving because there is not much friction (an outside

▶ THIS CROSS-COUNTRY SKIER MAY NOT KNOW IT, BUT NEWTON'S LAWS OF MOTION ARE AT WORK.

PHOTO: abkfenris/creativecommons.org

force) to work against you and bring you to a stop. (To make the friction even less, you can wax your skis.) To change direction, you have to push with both your poles and legs to turn your skis.

THE SECOND LAW

How hard do you have to push to get going? It depends on your mass—that is, the amount of matter in your body. The more you ate for lunch, the harder you'll have to push to get going quickly. This is a skier's way of using Newton's Second Law of Motion: The acceleration (rate of speeding up or slowing down) of an object depends upon the mass of the object and the force acting upon it.

You can start moving by pushing with a small force, but it will take you longer to reach the same speed than if you'd pushed harder. Regardless of how you choose to get going, what you're doing is putting to work some of that chemical potential energy from the food you ate at lunch and changing it into kinetic energy.

THE THIRD LAW

After you are in motion, you can coast for quite a long time, unless you hit another skier...or a tree! If you collide with another skier on the slopes, you may bounce off each other before coming to a stop. Newton had a law for that, too. It's his Third Law of Motion: For every action, there is an equal and opposite reaction.

READING SELECTION
EXTENDING YOUR KNOWLEDGE

▶ **SKILLFUL USE OF FORCES ENABLES A SKIER TO MAKE TRICKY MANEUVERS.**

PHOTO: Kevin Bernier/creativecommons.org

WHAT ELSE IS HAPPENING?

As you glide down those slopes, forces other than the ones you exert are helping you. One of them—you guessed it—is gravity. Gravity works in your favor when it comes to downhill skiing. In downhill skiing, you've got the whole mass of Earth helping you along.

The rate at which you speed up depends on a number of things, including the angle of the slope, how slippery your skis are, and another force—the friction created between your body and the air as you race along.

The smaller the amount of friction, the faster you'll accelerate. If you're wearing tight ski gear, you'll go even faster. The whole time you're going downhill, gravity is working to change your gravitational potential energy into kinetic energy.

What happens when you ski uphill? Gravity works against your upward motion, and you slow down. If the hill is higher than the one you started on, you won't make it to the top. If the second hill is lower than the one you started on, however, you'll have enough energy to go up and over the top and start down again.

When you're skiing, why can it be so hard to stop? One reason is that there's not much friction between the icy slopes and your well-waxed skis. Since there is not much friction, you just keep on going. If you want to stop, you have to apply a force against your motion. That means that, at least if you're like most first-time skiers, you fall when you're trying to come to a graceful stop.

At that moment, friction kicks in. The friction between your body and the snow slows you down. Your kinetic energy is converted to heat—so much heat, in fact, that as you slide, your body melts some of the snow! A more expert method of reducing your speed is to turn your skis so that they dig into the snow.

Would an understanding of the laws of motion, of gravity, and of friction have helped Sir Isaac be a better skier? It's hard to say. Skiing is one sport that requires a good deal of skill and athletic ability. But knowing what's happening as you glide along is a definite advantage. Sir Isaac

might not have had skills equal to those of an Olympic contender, but he probably would have managed to keep his balance, even when his wig was blowing in the wind. ■

▶ A DOWNHILL SKIER ENJOYS THE RIDE AS GRAVITY TRANSFORMS POTENTIAL ENERGY INTO KINETIC ENERGY.

PHOTO: Dylan Walters/ creativecommons.org

DISCUSSION QUESTIONS

1. How do Newton's Laws of Motion explain what happens during skiing?

2. In terms of energy and motion, in what ways is skiing like riding a roller coaster? In what ways is it different?

ASSESSING WHAT YOU'VE LEARNED

▶ **STUDENTS SETTING UP EQUIPMENT FOR THE ASSESSMENT.**

PHOTO: © Terry G. McCrea/ Smithsonian Institution

INTRODUCTION

In this unit, you investigated energy, forces, and motion. You explored how energy is transformed—how it changes from one form to another—and you examined the nature of different forces. You analyzed the motion of a fan car, a mousetrap car, and a roller coaster car. You also learned how to graph data and draw conclusions from your observations.

This lesson gives you a chance to show how well you learned the skills and concepts presented in the previous lessons. You will be given a set of data from an experiment. You are to analyze the data and then draw conclusions based on your analysis. In the second part of the assessment, you will complete short-answer and multiple-choice questions.

OBJECTIVES FOR THIS LESSON

Demonstrate understanding of principles and concepts of energy, forces, and motion.

Interpret data and draw conclusions based upon data.

▶ **MATERIALS FOR LESSON 13**

For you

1	copy of Inquiry Master 13.2: Multiple-Choice Questions
1	copy of Student Sheet 13.1: Data Analysis
1	copy of Student Sheet 13.2: Multiple-Choice and Short-Answer Response Sheet

GETTING STARTED

1. Review the objectives of this assessment with the class.

2. Listen as your teacher describes the assessment and its parts.

3. Make sure you have sharpened pencils and other materials your teacher says you need.

▶ **WHAT HAVE YOU LEARNED ABOUT ENERGY, FORCES, AND MOTION?**

PHOTO: Procsilas Moscas/creativecommons.org

DATA ANALYSIS

PROCEDURE

1 Go over Student Sheet 13.1: Data Analysis with the class. Ask questions about anything that isn't clear to you.

2 You will complete this section of the assessment on your own.

3 Your teacher will collect your student sheet when you have completed the data analysis.

MULTIPLE-CHOICE AND SHORT-RESPONSE QUESTIONS

PROCEDURE

1 Read the multiple-choice questions on Inquiry Master 13.2: Multiple-Choice Questions. Then, on Student Sheet 13.2: Multiple-Choice and Short-Answer Response Sheet, circle the choice that best answers each question or completes each statement. When you have completed the multiple-choice questions, respond to the short-answer questions on Student Sheet 13.2. Use complete sentences.

2 You will work independently on this part of the assessment.

3 Your teacher will tell you what to do with the student sheet when you have completed it.

REFLECTING ON WHAT YOU'VE DONE

1 Your teacher will show you a graph of the data you used in the data-analysis assessment. Think about the following questions:

A. How well did you follow good graphing procedures?

B. Were your conclusions based on evidence from the graph?

2 Ask questions and clarify your understanding of any multiple-choice questions or short-answer responses you answered incorrectly.

3 Read "Cars: Energy to Burn" to conclude the unit.

CARS
ENERGY TO BURN

During this unit, you've learned about energy transformations and force. How can you tie these ideas together into a single application? Take a ride in an automobile!

All cars, no matter how big or small, do two basic things. First, they transform chemical energy to heat energy and energy of motion. Second, they convert energy of motion to heat energy.

The heart of a car is its engine. In the engine, fuel (usually gasoline) burns in a cylinder, which is a closed container about the size of a 1-liter milk carton. As the fuel burns, it creates hot gases. These gases expand and press against the piston, which is a movable block at one end of the cylinder. Through a combination of rods, shafts, and gears, the piston is connected to the wheels of the car. The pressure of the expanding gases makes the piston move. This causes the wheels to rotate and makes the car move.

The moving car has kinetic energy, which is energy of motion. At this point, an important energy conversion has taken place: Some of the energy stored in the fuel has become energy of motion.

▶ **CARS RACE AROUND A TRACK.**

PHOTO: Martin Pettitt/creativecommons.org

But there's something else to think about. Not all the energy released when gasoline burns goes to help move the car. Most of it becomes heat energy. That is why the car needs a radiator. The radiator is the center of the car's cooling system. The cooling system circulates water and coolant around the engine to prevent it from overheating.

Once a car is moving, the next important energy transformation comes when it's time to stop. Friction helps here. Friction from the brakes stops the car; in other words, it reduces the car's kinetic energy. When the brake pads rub against the brake drums, the brakes get very hot—a sign that energy of movement is being transformed to heat energy. By the time that a car stops, most of the chemical energy that had been stored in the fuel tank has been converted to heat.

In a moving car, energy conversions are taking place continually. That's why it takes a constant supply of fuel to keep it going. When the fuel is used up, drivers need more. That means a stop at the fuel pump. Once the tank is filled, you have energy to burn! ■

DISCUSSION QUESTIONS

1. When you step on the brakes, what happens to the energy of the car? How does it happen?

2. What other machines that we encounter in our daily lives make energy conversions?

Appendix A
Directory of K'NEX® Parts

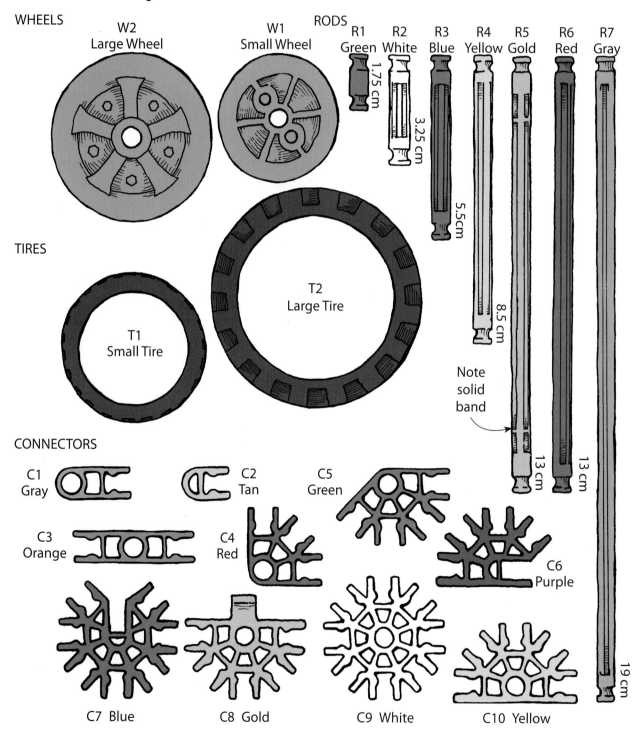

WHEELS

W2 Large Wheel

W1 Small Wheel

RODS
R1 Green
R2 White
R3 Blue
R4 Yellow
R5 Gold
R6 Red
R7 Gray

1.75 cm
3.25 cm
5.5cm
8.5 cm
13 cm
13 cm
19 cm

Note solid band

TIRES

T1 Small Tire

T2 Large Tire

CONNECTORS

C1 Gray
C2 Tan
C5 Green
C3 Orange
C4 Red
C6 Purple
C7 Blue
C8 Gold
C9 White
C10 Yellow

Appendix A
Tips for Assembling K'NEX® Parts

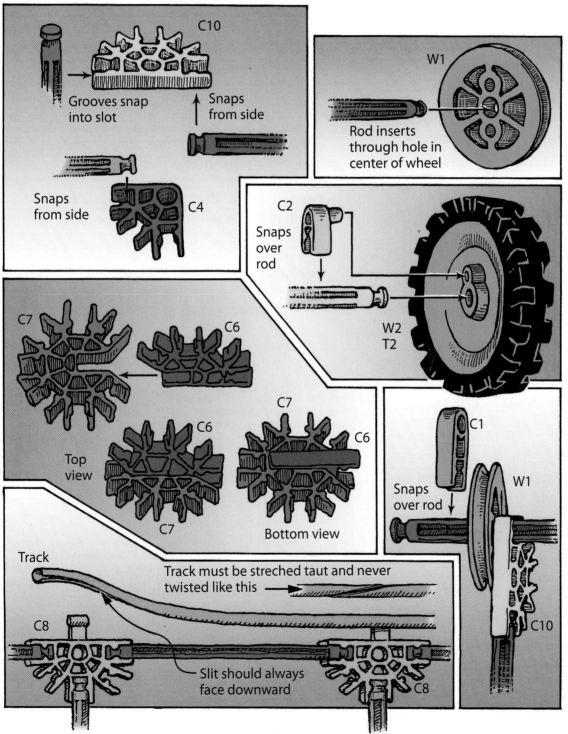

C10

Grooves snap into slot

Snaps from side

Snaps from side

C4

W1

Rod inserts through hole in center of wheel

C2

Snaps over rod

W2
T2

Pin inserts into hole in wheel to lock wheel and axle together

C7

C6

C6

C7

C7

C6

Top view

Bottom view

C1

Snaps over rod

W1

C10

Track

Track must be streched taut and never twisted like this

C8

Slit should always face downward

C8

Glossary

attract: To pull or draw near by physical force.

average speed: The distance an object travels divided by the time to travel the distance.

calibration: Aligning an instrument's scale with known values to ensure accurate readings.

constant: A condition that is not changed in a scientific experiment.

controlled experiment: A scientific investigation in which one variable is changed and all others are kept the same, or constant.

dependent variable: The variable in an experiment whose value is determined by the experiment. See also *independent variable.*

direct proportion: The relationship between two quantities in which the ratio of the quantities remains constant. In a direct proportion, when quantity A changes, quantity B changes by the same factor.

elastic force: The force exerted by elastic materials such as rubber bands and springs.

elastic material: A material that can be stretched or compressed and return to its original shape.

energy: The ability to do work. Energy exists in many forms, such as chemical, mechanical, electrical, thermal, nuclear, kinetic, and light.

experimental control: A variable or quantity that is not changed during an experimental procedure, while other quantities are changed.

experimental design: The process of planning an experiment or procedure to investigate a scientific question.

extrapolation: To make a prediction by extending a graph or data beyond the experimental values.

fair test: A controlled, scientific experiment.

force: A push or pull.

friction: A force that opposes the motion of an object. Examples include sliding friction and air friction. Sliding friction opposes the motion of objects across a surface. Air friction opposes the motion of objects moving through the air.

gravity: The force of attraction between all matter. Because of gravity, the earth attracts other objects and pulls them toward its center.

gravity assist: Using the gravity of a planet to change the motion (speed and direction) of a spacecraft.

heat energy: The energy that determines an object's temperature. Changes in an object's temperature or phase indicate a gain or loss of heat energy.

independent variable: The variable in an experiment that the experimenter changes. See also *dependent variable*.

inertia: Property of matter that makes matter resist a change in motion.

joule (J): Unit of work and energy in the metric system.

kilogram (kg): Unit of mass in the metric system.

kinetic energy: Energy associated with moving bodies. Kinetic energy is calculated by multiplying one-half by the mass of an object and by the square of its speed.

load or load force: The weight of an object that is lifted or carried.

mass: The measure of the amount of matter in a body. The mass of a body determines how much it will accelerate when unbalanced forces act on it and how much it weighs at the earth's surface. In the metric system, mass is measured in kilograms or grams.

magnet: An object made of iron, nickel, or cobalt or a combination of these materials and that has the ability to attract or repel other magnetic materials.

magnetic compass: A device that determines the presence and direction of a magnetic field.

magnetic field: A space or area in which magnetic forces can be observed. If a magnetic field exists in an area, magnetic forces will deflect a compass needle that is placed in that area.

motion: Displacement of an object over a period of time.

north-seeking pole: The end of a compass needle or magnet that points to the magnetic pole in the northern hemisphere of the earth if the needle or magnet is suspended freely.

net force: The sum of all forces acting on an object.

newton (N): Unit of force in the metric system.

permanent magnet: A magnet that does not lose it ability to attract iron, nickel, and cobalt as time passes.

potential energy: Stored energy that can be released; includes chemical, electrical, gravitational, and nuclear energy.

repulsion: To push away, the opposite of attract.

south-seeking pole: The end of a compass needle or magnet that would point to the earth's magnetic pole in the southern hemisphere if the needle or magnet was suspended freely.

spring scale: A calibrated spring used to measure forces.

speed: A measure of how fast something is moving. The average speed of an object over a distance is the distance it moves divided by the time it takes to move that distance.

surface area: The measure of the total surface of an object. For a block, the surface area of any side of the block is the side's length multiplied by its width.

technological design: The process of designing solutions and building devices to meet human needs.

variable: An element in an experiment that can be changed.

weight: The measure of the force of gravity on an object. Weight can be measured in newtons or pounds.

work: The product of a force and the distance through which it acts.

Index

Photo Credits

Front Cover
NPS Photo by Jim Peaco

Lessons
2 ©Terry G. McCrea/Smithsonian Institution
4 chispita_666/creativecommons.org **9**
Library of Congress, Prints & Photographs
Division, LC-USZ62-47604 **10 (top)** Courtesy
of Smithsonian Institution Libraries,
Dibner Library of the History of Science
and Technology, Washington, D.C. **(bottom)**
NASA Jet Propulsion Laboratory **12** U.S.
Navy photo by Photographer's Mate 2nd
Class Isaiah Sellers III **14** National Science
Resources Center **18 (top)** Library of
Congress, Prints & Photographs Division,
LC-USZ62-10191 **(bottom)** Library of
Congress, Prints & Photographs Division,
LC-USZ62-44624 **20** Wikimedia Commons
22 Mykl Roventine/creativecommons.
org **26** Russell Neches/creativecommons.
org **27 (top right)** ©Mark Karrass/Corbis
(bottom left) lpiepiora/creativecommons.
org **28** jmagnusphoto/creativecommons.
org **30** National Science Resources Center **31**
©2009 Carolina Biological Supply Company
32 ©2009 Carolina Biological Supply
Company **34** ©2009 Carolina Biological
Supply Company **36** U.S. Air Force photo
by Staff Sgt. Matthew Hannen **38 (top)** CJ
Smithson/creativecommons.org **(bottom)**
NASA Spinoff Magazine **39** U.S. Navy
photo by Mass Communication Specialist
2nd Class Christopher Stephens **40** Iwona
Erskine-Kellie/creativecommons.org **41**
(left) Courtesy of Black Diamond Equipment
(right) Cristian Ordenes/creativecommons.

org **42** reggestraat/creativecommons.org
44 Juan Tello/creativecommons.org **47**
(top) ©2009 Carolina Biological Supply
Company **(bottom left)** ©2009 Carolina
Biological Supply Company **(bottom right)**
©2009 Carolina Biological Supply Company
48 Chip Clark, National Museum of Natural
History, Smithsonian Institution **49** NOAA
Photo Library, NOAA Central Library, OAR/
ERL/National Severe Storms Laboratory
(NSSL) **51** Courtesy of Los Alamos National
Laboratory **52** ©Roy McMahon/Corbis **60** H.
Powers/U.S. Geological Survey **61** Richard S.
Fiske, National Museum of Natural History,
Smithsonian Institution **64** ©Exploratorium,
www.exploratorium.edu **66 (top)** ©National
Maritime Museum, Greenwich, London
(bottom) National Science Resources
Center **67** mill56/creativecommons.
org **68** Leonard Low/creativecommons.
org **70** Rich Moffitt/creativecommons.org
72 National Science Resources Center **77**
NASA/Johns Hopkins University Applied
Physics Laboratory/Southwest Research
Institute **78-79** NASA/Johns Hopkins
University Applied Physics Laboratory/
Southwest Research Institute **80** NASA/
Johns Hopkins University Applied Physics
Laboratory/Southwest Research Institute
82 (top) pointnshoot/creativecommons.org
(bottom) ©Zamboni Company, David Klutho
83 Shea Hazarian/creativecommons.org **84**
Universal Hovercraft, www.hovercraft.com
86 National Science Resources Center **90**
Marvin D. Blimline, Maryland State Highway
Administration **91** Marvin D. Blimline,
Maryland State Highway Administration